Sedona Through Time

Wayne Ranney

SEDONA
THROUGH TIME

A Guide to Sedona's Geology

Wayne Ranney

Illustrations by Bronze Black
Paleo-maps by Ron Blakey

SEDONA THROUGH TIME
A Guide to Sedona's Geology
3rd Edition
Fully Revised and Updated, January, 2010; 2nd printing, June, 2013

Text by Wayne Ranney
Photography by Wayne Ranney and Bronze Black
Paleogeographic maps by Dr. Ron Blakey
Edited by Pam Frazier
Maps and diagram illustrations by Bronze Black
Book design by Bronze Black
Printed in China
Printed and bound by P.Chan & Edward, Inc.

Reprint History 12 11 10 9 8 7 6 5 4 3 2

ISBN 978-0-9701203-8-0
Library of Congress Control Number: 2008943811
Third Edition 2010

Wayne Ranney
255 E Hutcheson Dr.
Flagstaff, Arizona 86001
www.wayneranney.com

Cover photograph: © Bronze Black
Back cover photographs–top: © Bronze Black; middle and bottom: © Wayne Ranney
Half title photographs: © Bronze Black
Full title photograph: © Wayne Ranney

Table of Contents

List of Illustrations

Bell Rock near the Village of Oak Creek

Preface and Acknowledgments

S *edona Through Time* made its first appearance in bookstores in 1993. Since then it has been revised once and reprinted many times. It is heartening to look back on the history of this little book and reflect on its many successes. And while Sedona's residents and jeep tour guides still find its story interesting and compelling, I realize that the way the story has been told is a bit dated. Another edition of this popular book is in order, not so much because the geology has changed, but because of improvements in the way this information can be shared. It is my pleasure to present this fully revised and updated third edition of *Sedona Through Time.*

Since those earlier editions I have progressed steadily in my ability to observe, teach, and write about the geologic phenomena in and around Sedona and across the globe. The way in which I tell geologic stories and the topics that I choose to talk about have also changed. In this new edition I have decided to include alternative interpretations for how parts of Sedona's landscape may have evolved. These ideas are favored by some local geologists who don't necessarily see things the same way I do. Nevertheless, I want to give voice to these alternative views and make readers aware of all the possible ways in which the landscape in Sedona could have evolved. The reader may then come to know that not all of area's landscape puzzles have been fully deciphered.

The first edition of this book came into existence through the efforts of Larry Stevens, Gwen Waring, Nancy Nelson, and Jeanette Smith. Many other people contributed to its success as well. Thanks go to my geology professors at Northern Arizona University, who taught me how to see the earth through time. Among this group I give special thanks to Dr. Ron Blakey who passionately shared his earliest thoughts about the origins of the red rocks with me and other geology students. Dr. Richard Holm shared his broad knowledge of volcanic rocks, and Dr. Dale Nations directed me towards a seemingly insignificant volcano called House Mountain. Two other fine local geologists are also worthy of mention here—Paul Lindberg and Stan Beus—both of whom are great observers and interpreters of the landscape and continue to share their geologic insights with all of us.

This edition is completely redesigned through the artful talents of Bronze Black. He also drew the location maps, the cross sections, and the diagrams. Pam Frazier edited the text and her diligence, poise, and knowledge in these matters is much appreciated and makes the book much better. Ron Blakey contributed the splendid paleogeographic maps that grace these pages. Readers will undoubtedly be impressed with the detail of these beautiful and informative maps. I include new and updated road logs in this edition that replace the original thirty-mile log that described the scenery from the top of the switchbacks in Oak Creek Canyon to Interstate 17. At the suggestion of many, I have added geology-themed hikes to selected localities in and around Sedona. These geologic hiking guides were influenced by the previous work of Cora Ruhr and Gail Theilacker and I am especially indebted to them.

Through the years, I have developed very gratifying personal friendships with people who share their love of the red rocks with me. Among these are Brenda Robinson, George Abbott, Clint and Louise Gelotte, Sandy Unger, Fred Johnson, Peter Pilles, Jack and Marlene Conklin, Chris Weld, Bryce Babcock, and Jean Kindig. To all of my former and present students at Yavapai and Coconino community colleges I applaud your enthusiasm for learning and give thanks to you for your encouragement to me to share Sedona's geology with a wider audience. Thanks also to the many jeep tour guides in Sedona who have constantly endeavored to improve their stories so that they may share them in more meaningful and appropriate ways with the countless visitors they take into red-rock country every year.

Special thanks must go to my family and friends, who continue to support me in my unconventional career. To my mother, Prudence Ricca Ranney, who at this time is traveling through some far-off, beautiful landscape in the heavens, I thank you for an exceptionally loving and grounded foundation. To my dad, Donald Ranney, I give special thanks for bringing home those fossils when I was a boy. And lastly, I must give thanks to Helen my wife, who is the ultimate inspiration for this third edition. Without her encouragement and support, this book would have gone out of print. Thank you Helen!

My ultimate goal has always been to share complex geologic information with a wide and varied audience. It is something that is important to me and I thank all of you for your support in my endeavor to accomplish this. But most of all I thank you just for listening to these stories of *Sedona Through Time*.

View from Bell Rock

Ancient ruins among Sedona's red rocks

Blooming agave

Introduction

The red rocks exposed near Sedona, Arizona, reveal a fascinating story from a time long gone on planet Earth. Rocks 316 million years old have been laid bare by erosional events that began at least 80 million years ago. But ironically, it has been only in the last 200 years or so that humans have come to understand that the history of our planet is contained within the earth's rock **strata** (words appearing in **bold** are defined in the glossary beginning on page 152). It's remarkable to think that a person with some basic college instruction in geologic principles can look at layers of ordinary **sandstone** or **limestone** and resurrect an ancient landscape hundreds of millions of years old. Incredibly, our species has learned to see the earth through time. We have become time-travelers with the ability to see and experience the earth's ancient landscapes, as well as its ancient life forms and ecosystems.

We are among the earliest generation of people to see our earth through this hazy veil of time, to bear witness to its dynamic change, to feel and sense its pulse. This modern geologic awareness propels us into new ways of looking at our planet and forces us into a whole new relationship with it. The new awareness that we are only a recent arrival in this long-term spectacle of Earth history is a great lesson in humility, and our more expansive view of Earth history is as much a philosophical or spiritual revelation as it is a scientific one. This innovation in our thinking is perhaps as important a milestone in human history as the domestication of fire or the invention of the telescope. This is an exciting time to be a part of this new way of seeing our place upon this planet.

With its stunning red-rock scenery, Sedona is a spectacular place to observe and experience a portion of this grand pageantry. It is one of the top tourist destinations in all of Arizona, with approximately four million people visiting the area each year, mostly because of the dramatic beauty of the landscape or to enjoy some cultural or recreational amenity within sight of it. Some people might mistakenly believe that this immense popular interest has spilled over into the area of geologic research and that all there is to know about the geology of the red rocks has been determined. In a place as beautiful and interesting as Sedona, this would seem only natural. But within the borders of a state that boasts the Grand Canyon, Monument Valley, and the Petrified Forest, the Sedona area remained somewhat hidden from view, tucked beneath the forested **Mogollon Rim**.

The History of Rock Names in Sedona

In the first half of the twentieth century the rocks near Sedona were thought to be a near identical replication of those found at the Grand Canyon, where, by virtue of its impressive size and unequaled exposure, much of Arizona's strata were first described.

Geologists took the names of the rock **formations** found at the Grand Canyon and extended them quite naturally into the Sedona area, knowing that the flat-lying sedimentary rocks, although hidden beneath the surface, are continuous across the expanse of the Coconino Plateau.

It wasn't until 1945 that the eminent Grand Canyon naturalist and geologist Edwin "Eddie" McKee first proposed specific formation names for the rock strata found in Oak Creek Canyon near the tiny village of Sedona. Since McKee had described and named some of the strata in the Grand Canyon, he readily applied the same **nomenclature** to Oak Creek Canyon. He classified all of the red rocks around

OLD TERMINOLOGY

SUPAI C

SUPAI B

SUPAI A

Sedona as belonging to the Supai Formation of **Pennsylvanian** and **Permian** age, 316 to 251 **Ma** (mega-annum, meaning millions of years ago). McKee further proposed that the Supai be subdivided into three informal units or **members**, beginning with the lowermost A member, the middle B member, and an upper C member. Geologists were comfortable with his classification scheme and accepted it, noting only that the Hermit Formation found at the Grand Canyon was missing in Sedona. It's still somewhat surprising that the famous red rocks did not have any geologic classification prior to 1945.

In 1979, Ron Blakey, a young professor from Northern Arizona University in Flagstaff, began looking at the red rocks in Sedona near Schnebly Hill. Like McKee, Dr. Blakey recognized a tri-part subdivision of the red rocks but thought that instead of all three belonging to the Supai Formation, they might belong to three different formations, each of which deserved a specific name. Under Blakey's proposal, McKee's old Supai A member was recognized as the Supai Group (meaning that it

NEW TERMINOLOGY

SCHNEBLY HILL FORMATION

HERMIT FORMATION

SUPAI GROUP

© Bronze Black

was actually a package of four different formations), the old Supai B member was the Hermit Formation, and the old Supai C member became a newly described unit not present in the Grand Canyon at all, which Blakey called the Schnebly Hill Formation.

Dr. Blakey's classification scheme soon came under scrutiny by other geologists who saw things a little differently and an intellectual debate of sorts commenced.

Geologic Time and Significant Geological Events

Time	Geologic Eon	Era	Period	Rocks of the Sedona Area	Significant Geological Events in N. Arizona	Other notes
0		Cenozoic Era	Neogene	Oak Creek volcanics / Verde Formation / Hickey Formation / Beavertail Butte Formation	Carving of Oak Creek Canyon / Verde lake	Age of the Mammals
23			Paleogene	Rim gravel	Formation of the Mogollon Rim	
65		Mesozoic Era	Cretaceous		Uplift of the Mogollon Highlands	Age of the Dinosaurs and the Cycads
144	Phanerozoic Eon		Jurassic	(Mesozoic rocks have been eroded from the Sedona area)		
200			Triassic		Break-up of Pangea	
250		Paleozoic Era	Permian	Kaibab Limestone / Toroweap Formation / Coconino Sandstone / Schnebly Hill Formation / Hermit Formation	Pangean Supercontinent	Age of the Fish, and Ancient Life
287			Pennsylvanian (Carboniferous)	Supai Group		
320			Mississippian (Carboniferous)	Redwall Limestone	Tropical Seas	
360			Devonian	Martin Formation		
408			Silurian			
438			Ordovician			
505			Cambrian	Tapeats Sandstone	Sandy Shoreline	
540						
1.0 (Billions of Years ago)	Proterozoic Eon	Precambrian (88% of Earth history)				Multi-celled Life
1.6				Precambrian Crystalline Rocks	Submarine Volcanoes near Jerome	Single-celled Life
2.5						
3.8	Archean Eon			No Rock Record		No Rock Record
4.6					Earth Forms	

Millions of Years ago

This was most confusing to the residents of Sedona who hadn't noticed any change whatsoever in the way the red rocks looked. They were accustomed to using McKee's classification, calling all red rocks in Sedona the Supai Formation. I remember well the confusion that reigned in the early days of my career as a geologist. People didn't know what to call the rocks, even though they hadn't changed one bit.

Time, however, has been kind to Dr. Blakey's classification scheme. As he and his graduate students "went to the rocks" to determine exactly what they were, they discovered the proper relationships that allow for the **correlation** of strata at Sedona with other areas of the **Colorado Plateau**. These studies provide evidence of the different environmental conditions that once existed here. Without these studies, much of the information contained in this book would still be hiding in the cliffs and canyons of red-rock country. Dr. Blakey's terminology for the strata has become the accepted language used around Sedona. Incredibly, this area preserves a rock formation (the Schnebly Hill Formation) that the Grand Canyon does not have.

Geologic Time in Sedona

Sedona's geologic story begins about 316 Ma, an amount of time so immense that it is virtually impossible to comprehend. None of the earth's present-day landscapes existed then and many different landscapes literally have come and gone since that time. Even if we could go back only 20 million years, we would not recognize much of the modern landscape seen near Sedona today. With some knowledge of the area's modern geography, one might be able to recognize certain rock strata by its color or erosional form, but none of today's specific landforms was in existence then. If we are to make sense of what the rocks here are telling us, we will have to become familiar with the concept of geologic or deep time.

Not even geologists truly comprehend the vast length of time that geology represents, but a few analogies may help. Think of all the seconds that tick away every day of our lives. In just one day, 86,400 seconds pass by. It takes about ten years for 316 million seconds to pass. That many minutes would take us back to the time when prehistoric farmers were growing corn in the Verde Valley (about AD 1410). Another way to understand how incredibly large these numbers are, is to imagine tiny sand grains stretched out in a row. Each sand grain is only 1/25 of an inch in length but 316 million of them laid end to end in a straight line would stretch almost 200 miles. Geologic time is truly immense and readers must apply all of their intellect and a great deal of imagination to understand how the Sedona area could look so different at so many different time periods in the geologic past.

The area around Sedona appeared vastly different 316 million years ago. It was a low, featureless **floodplain** where many different layers of sediment would accumulate, intermittently, during a 46-million-year period. We say "intermittently" because sediment was not accumulating continuously during this time. At times, the area was elevated just enough so that sediment could not accumulate or perhaps was even being removed by erosion. We call these gaps in the rock record unconformities, and although there is no rock to read at an **unconformity**, its very absence tells us something about the uplift history of the area we are studying. Every bit of information is helpful in reconstructing the past, even **horizons** known as unconformities.

Geologists are still busy assimilating all of the information contained in Sedona's red rocks. Certainly future studies will dig up even more secrets from within this remarkable area and this book will not be the final word with respect to some of the specific periods of history discussed herein. This is especially true for the interpretations that have been made for the river gravels that were deposited in this area between about 50 and 20 million years ago. There are different, and perhaps conflicting, ideas regarding these deposits, making an interpretation of the landscape's appearance at this time rather uncertain.

Still, our understanding of the area now has a basic framework and the overall story of red-rock country is generally known. Prepare yourself for a fantastic journey back in time, when northern Arizona was, in turns, a coastal plain, a shallow sea, a sandy desert, and the site of an ancient volcano. You will become a time traveler, capable of moving backward or forward through a plethora of wonderful landscapes. The rocks, and hopefully this guidebook, will reveal to you what we know about this awe-inspiring landscape. My attempt to reconstruct just one of these past landscapes (the House Mountain volcano), forever changed my relationship with our planet. The sense of discovery I experienced while digging out this little secret of Earth history is the one thing that I especially hope to share with you, the interested reader. This book is written to stimulate your intellect and satisfy your curiosity about this remarkable place. The geologic story of Sedona's red rocks is in your hands!

© Bronze Black

Sedona's Bedrock Stratigraphy

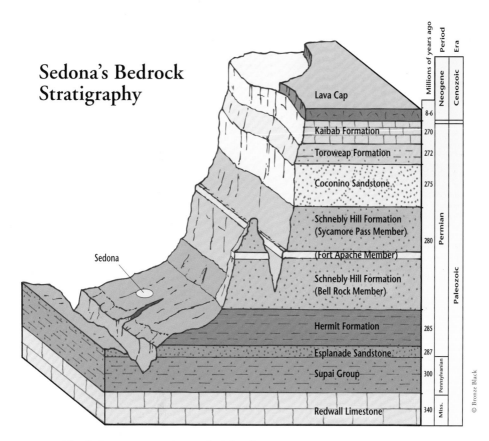

	Millions of years ago	Period	Era
Lava Cap	8-6	Neogene	Cenozoic
Kaibab Formation	270	Permian	Paleozoic
Toroweap Formation	272		
Coconino Sandstone	275		
Schnebly Hill Formation (Sycamore Pass Member)	280		
(Fort Apache Member)			
Schnebly Hill Formation (Bell Rock Member)			
Hermit Formation	285		
Esplanade Sandstone	287		
Supai Group	300	Pennsylvanian	
Redwall Limestone	340	Miss.	

Sedona

© Bronze Black

The Sedona area contains a colorful and interesting sequence of strata capped by relatively young volcanic lava flows. Refer back to this diagram often to become familiar with this sequence.

View to the northeast from Cathedral Rock

View from Bear Mountain

PART 1

Geography and Setting
of the

Red Rocks

© Bronze Black

Geography and Setting

T his story takes place among the fantastic red rocks exposed near Sedona, Arizona, a small city with a population of about 15,000 people. Sedona is located in north-central Arizona about 120 miles north of Phoenix and about 25 miles south of Flagstaff. The city straddles the boundary between two geologic **province**s, the Colorado Plateau to the north and the **Transition Zone** to the south.

The Colorado Plateau is a land of prominent flat-topped mesas, graceful sandstone cliffs, and deep, spectacular canyons. The plateau country has mostly escaped the occasional periods of squeezing and stretching that other nearby regions have

© Wayne Ranney

experienced and this is why the strata here are, for the most part, still very near their original horizontal position. The Colorado River bisects the heart of this red-rock wonderland and numerous national parks, monuments, forests, and recreation areas are found within its drainage basin.

The Transition Zone is a small but important province that intervenes between the northern plateau country and the **Basin and Range** Province in central and southern Arizona. The Basin and Range is a land of low desert valleys separated by rugged mountains that rise like islands above the surrounding valleys. The Basin and Range is a young province that began to form about 17 Ma when the earth's crust was stretched and thinned, forming the valleys in which Phoenix and Tucson are located today. The Transition Zone displays characteristics common to both the Colorado Plateau and Basin and Range provinces—it contains rock strata like that

View from the top of Airport Mesa reveals Sedona's spectacular setting

of the Colorado Plateau but has been stretched and broken like the Basin and Range. Cottonwood, Camp Verde, Prescott, and Payson are some of the cities located within the Transition Zone.

The dividing line between the two provinces is the **Mogollon Rim**, a high, forested **escarpment** (cliff) that separates the low cactus deserts in the south from the pine-studded plateaus to the north. The Mogollon Rim was named after Juan Ignacio Flores Mogollón, the Spanish governor of New Mexico from 1712 to 1715. During his administration, exploration parties were dispatched to survey for any mineral wealth that might be found in the area. A major silver lode was discovered in the Mogollon Mountains in present-day New Mexico. The name Mogollon Rim has historically been used for the escarpment located just north of Payson, Arizona, and a lot of early Arizona history and western lore is associated with the area. Zane Grey built a cabin there and set some of his famous novels in and around this rugged country. Geologists, however, recognize that the quintessential Mogollon Rim near Payson is related to and continuous with the red-rock country near Sedona. They have shown that the rocks in these two areas share a similar geologic development and the term Mogollon Rim is widely used in describing the red rock escarpment near Sedona. The rim plays a starring role in the more recent story of how the modern landscape came to be.

The climate of the area is variable because of the great change in elevation from one place to the next. The city center is located at about 4,500 feet above sea level, and all seasons are mild compared to the northernmost or southern parts of the state. Sedona usually receives a few dustings of snow in winter but the sun is never hidden for long and photographers must be up early in the morning to capture the stunning contrast of red rocks capped with snow. Meanwhile, snow may keep roads closed on the Mogollon Rim in the Coconino National Forest well into spring. This winter snow brings only half of the eighteen inches of annual precipitation to the area. The other half comes in torrential summer rains. This rain comes from spectacular thunderstorms or cloudbursts that quickly fill the area's streams and rivers with their silt-laden runoff. Summers can be hot here, but not nearly as hot as Phoenix or Tucson.

Oak Creek is the major drainage in the Sedona area and is born from springs that issue from beneath the Mogollon Rim and within Oak Creek Canyon. This stream flows south through the canyon to the city of Sedona, then turns west to flow around the north side of House Mountain, and bends south again through Page Springs and Cornville before merging with the Verde River. The groundwater hydrology of Oak Creek is of extreme importance to the city of Sedona. Other nearby streams are (from west to east) Sycamore Creek, Dry Creek, Secret Canyon, and Dry Beaver and Wet Beaver creeks. All originate beneath the Mogollon Rim and are tributaries to the Verde River.

A brief human history of the area is certainly in order. It is unknown when the first person saw the red rocks but certainly the mammoth hunters did so 13,000 years ago, though the evidence for their presence is sparse. After the mammoth's demise, other hunters and gatherers found the area's abundant game to their liking and these traditions held on in the Verde Valley longer than in other nearby localities.

Geologic Provinces of the Southwest

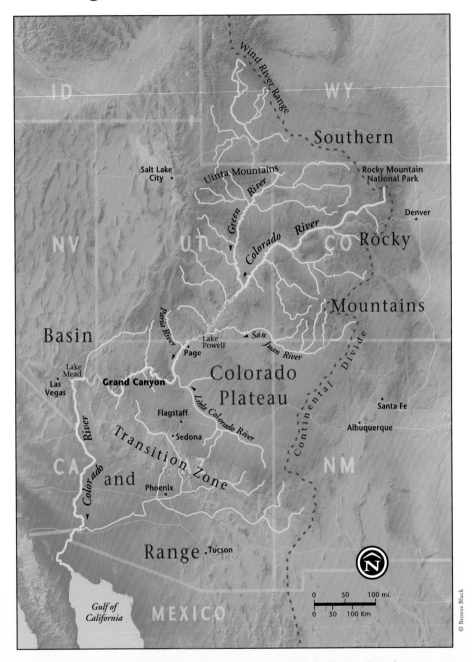

Map showing the geologic provinces that play a major role in the creation of Sedona's landscape. The city of Sedona is located at the boundary of the Colorado Plateau and Arizona's Transition Zone.

By AD 600 or 700 a more sedentary lifestyle was adopted and these people, called the Sinagua by archaeologists, lived in pit houses and emphasized agriculture in their societies. The classic period of the Sinagua lasted from about AD 1130 to AD 1425, and many well-known ruins and cliff-dwellings in the area, such as Palatki, Honanki, Tuzigoot, and Montezuma Castle, were built and inhabited at this time. The architectural and material remains of the Sinagua show close affinities with the historic Hopi Indians of northeastern Arizona. Oral traditions among the Hopi mention a place called Palatkwapi, "the red land to the south." A link between the Hopi and the red-rock country seems secure, even though the people were gone from here by AD 1400 or a little later.

However, before the Sinagua abandoned the Verde Valley for points north, another group, the Wipukpaia (Yavapai), may have inhabited the valley alongside them. By about AD 1500 the Dil zhe'e' (Tonto Apaches) shared the land with the Yavapai and intermingled with them for certain cultural activities, although the two groups remained culturally distinct. In the 1870s, General George Crook confined both groups to a reservation in the valley, but increased Anglo expansion associated with farming and mining led to the abolishment of this reservation. The people were forcibly removed to the San Carlos Reservation in east-central Arizona. In the year 1900, the United States government turned a blind eye when the people returned to their home in the Verde Valley. Numerous small reservations of the officially recognized Yavapai-Apache Nation now dot the area today.

The first non-Indian settler in the area was John James (Jim) Thompson who arrived in 1876. By the turn of the 20th century, about fifteen families lived in or near the mouth of Oak Creek Canyon. In October, 1901, T. C. Schnebly arrived from Missouri to the mining town of Jerome and was joined two weeks later by his wife, Sedona, and their two children. They soon settled along the banks of Oak Creek with the intent of supplying fresh vegetables to residents of Jerome. In 1902 T. C. applied to the government for a post office and his brother suggested the name Sedona. In spite of the area's dramatic beauty, the small settlement contained fewer than 500 people until the late 1940s, when Hollywood, in the throes of a Western film binge, discovered the spectacular setting. Soon Arizona Highways and other magazines featured the red rocks in full-color articles, extolling the magic of the area and its salubrious climate. Real estate became the number one industry in Sedona during the 1960s and 1970s, in spite of numerous attempts by some local residents to declare portions of the area a national monument. In 1984, many of the canyons along the Mogollon Rim were designated as official wilderness areas, ensuring that natural beauty will forever be the area's greatest asset. Sedona was incorporated in 1988.

Residents are proud of their red-rock heritage even if it has come at the cost of their small-town ambiance. Living in Sedona invokes feelings of joy that few places can match and many people find the place spiritually enriching. It is only fitting then that residents and visitors alike become more aware of the fascinating geologic story that the area holds. When people come to know a place deeply, it not only enriches their lives, it also inspires them to work with pleasure to take care of it.

Map of the
Sedona Area

Switchbacks

To Flagstaff

West Fork

Oak Creek

Page

Flagstaff

Sedona

Phoenix

Tucson

Mogollon Rim

Bear Mountain

Sliderock State Park

Oak Creek Canyon

Wilson Mountain

89A

Oak Creek

Legend

I-17	Interstate Hwy
89A 179	Highway
	Road
	Road log 1
	Road log 2
	Creek
	Dry Wash

Doe Mesa

Capitol Butte

Sedona

Airport Mesa

Oak Creek

Munds Mountain

Lee Mountain

89A

Dry Creek

Cathedral Rock

Red Rock State Park

179

Bell Rock

Courthouse Butte

Village of Oak Creek

Horse Mesa

SCALE

Miles
0 1 2 4

Km
0 2 4

N

House Mountain Volcano

Beavertail Butte

I-17

Page Springs

Page Springs Road

Beaverhead Flat Road

Beaverhead

Cottonwood

Verde River

Cornville

Cornville Road

179

260

© Bronze Black

To Phoenix

Red rock view near Uptown Sedona

PART 2

Forming the

Red Rocks

© Bronze Black

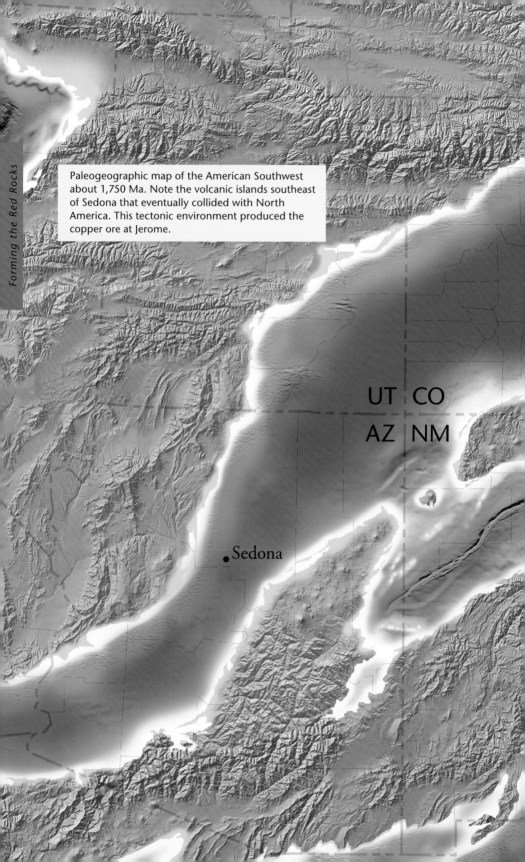

Paleogeographic map of the American Southwest about 1,750 Ma. Note the volcanic islands southeast of Sedona that eventually collided with North America. This tectonic environment produced the copper ore at Jerome.

UT CO
AZ NM

•Sedona

Prelude to the Red Rocks:
SUBMARINE VOLCANOES, BEACHES, AND TROPICAL SEAS
Before 316 Million Years Ago

Although very few people may wonder what lies beneath Sedona, there are rocks not yet exposed that can reveal what occurred here long before the red rocks were even a gleam in Mother Nature's eyes. These rocks are lying in wait for their day in the sun when the crimson strata will be stripped away. These older rocks are already exposed high in the Black Hills near Jerome and tucked away in the recesses of the Verde River Canyon where the train goes to Perkinsville. These rocks have been studied and much of their history is known.

The oldest rocks in the Verde Valley, and in Arizona for that matter, display evidence for the presence of great submarine volcanoes that spewed lava on an ancient sea floor. It was within this setting that the world-class copper ores of Jerome were emplaced. Imagine 1,750 Ma when the future area of the Verde Valley lay near the southern coast of ancient North America. A chain of volcanic islands, similar to modern-day Japan or Indonesia, was drifting northwest towards the continent, infusing the area with the volcanic energy that ultimately created these rocks. Catastrophic **submarine** eruptions emplaced the Cleopatra crystal **tuff**, and the volcanic piping system that rose through it left large quantities of copper, gold, zinc, and lead. The eventual collision of the island volcanoes with North America severely deformed the volcanic rocks into a maze of confusing patterns that puzzled the old miners for decades. These rich ores lay completely hidden by younger rocks until movement on the Verde Fault sliced the landscape open about 10 million years ago. More on this subject later.

A tremendously long period of erosion then followed the volcanic excitement. Huge thicknesses of sediment once may have covered the eroded remains of the deformed volcanic pile; evidence from the Grand Canyon suggests this possibility. But even if this were true here, all of the evidence is now gone from the Verde Valley. A gap in the rock record, called The Great Unconformity, is all that remains to represent more than 1,200 million years of time. This figure boggles the mind.

Erosion ultimately planed a low surface that was inundated by a sea by about 525 Ma. This sea entered the region from what appears to be the west today but was to the north all those millions of years ago. Along this shoreline, which stretched from Mexico to Canada, coarse sand was laid down in beach environments. Today we call this the Tapeats Sandstone, and this durable brown rock was quarried from outcrops in the Black Hills and used as a building stone in many historic structures in Jerome. It was even used in some curbside gutters in the town center. Other sediments may have covered the area soon after the Tapeats was deposited (again the evidence comes from the Grand Canyon), but these layers are now gone. Another period of erosion ensued, lasting perhaps 150 million years, and on top of this the Martin Formation was laid down in a warm, shallow **seaway**. This limestone is dated at about 375 Ma and contains a few fossils of corals and mollusks.

By about 340 Ma, the area of the future Verde Valley was well below sea level with the equator nearby to the south. California and Nevada did not yet exist as emergent land areas and the foundations of those states were just beginning to form. This was a time when Arizona was the floor of a clear, tropical seaway. The shells of various **marine** animals (crinoids, bryozoans, brachiopods, nautiluses, and corals) can be found within these deposits, called the Redwall Limestone. Exposures of this lime-stone can be seen in and around Jerome where **faulting** has sliced it up and placed it both above and below the town limits.

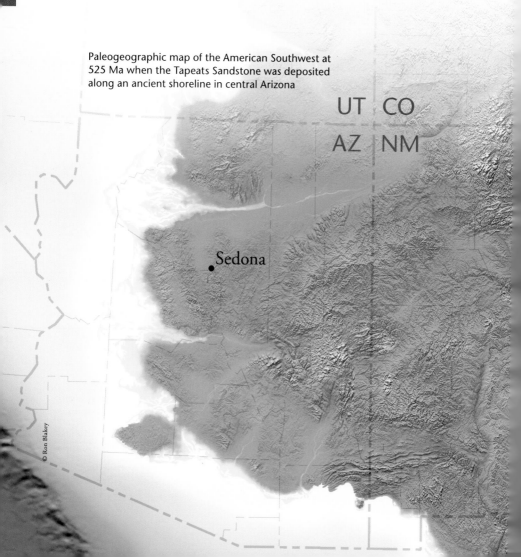

Paleogeographic map of the American Southwest at 525 Ma when the Tapeats Sandstone was deposited along an ancient shoreline in central Arizona

UT CO

AZ NM

Sedona

© Ron Blakey

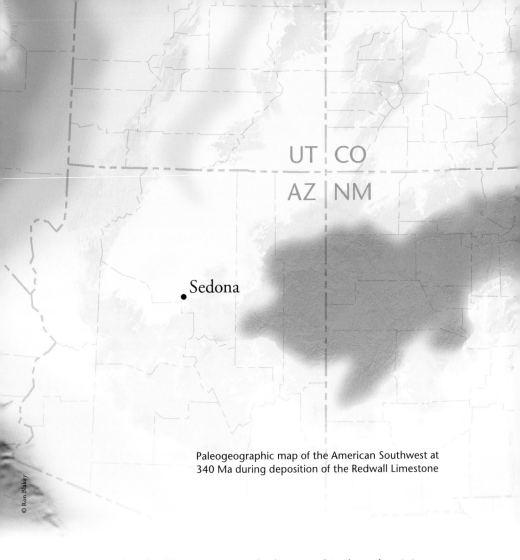

UT | CO
AZ | NM

.Sedona

© Ron Blakey

Paleogeographic map of the American Southwest at
340 Ma during deposition of the Redwall Limestone

The Martin and Redwall limestones were both quarried in the early mining days and used to make cement mortar in the construction of buildings in Jerome. In the late 1950s, Clarkdale became the home of Salt River Materials (then called the Phoenix Cement Plant) because all of the ingredients needed for making high-grade cement could be found within the calcium-rich (Redwall) and silica-rich (Martin) limestone. All of the cement used in the construction of Glen Canyon Dam (which impounds the waters of Lake Powell) was hauled to Page, Arizona, from this quarry in the Verde Valley. It is remarkable to think that strata from these ancient tropical seas was crushed, cooked at high temperatures, and transported by truck up and over the Mogollon Rim in the building of this massive dam.

After about 20 million years the sea **retreated** from the area near Sedona and a low coastal floodplain developed on top of the Redwall Limestone. In a very short time conditions once again would be conducive to the accumulation of sediment, this time leaving behind the famous red-rock formations in their wake.

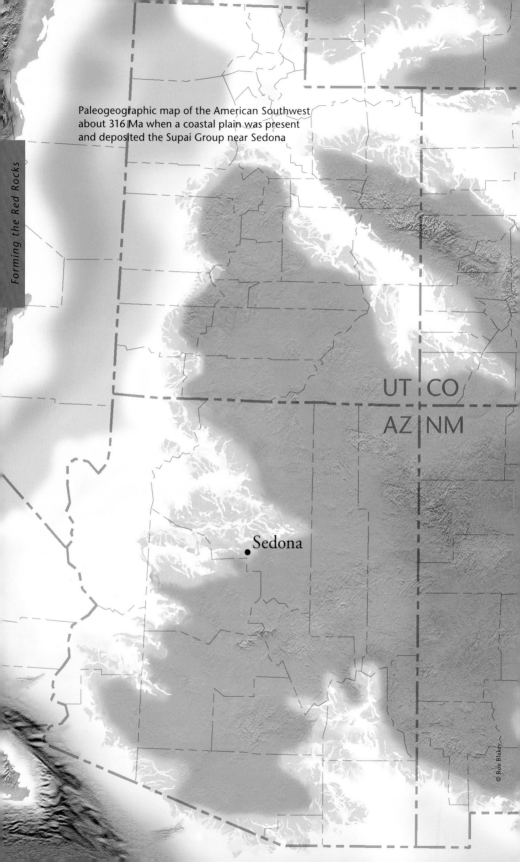

Paleogeographic map of the American Southwest about 316 Ma when a coastal plain was present and deposited the Supai Group near Sedona

Forming the Red Rocks

UT   CO
AZ   NM

● Sedona

© Ron Blakey

The Supai Group:
A COASTAL PLAIN
316 to 287 Million Years Ago

The lowest layer (and thus the oldest) of Sedona's red rocks is called the Supai Group. It is mostly buried by other layers and not readily visible within the city. It is partially exposed beneath Midgley Bridge in Oak Creek Canyon and its entire 600-foot thickness is exposed in the heart of the Sycamore Canyon Wilderness west of Sedona. It is composed of alternating layers of sandstone and mudstone that form many stair-step cliffs and slopes of red strata. As clay-size particles from the red-colored Supai wash across the underlying gray limestone, they stain the surface of the limestone, giving the Redwall Limestone its color and name.

The Supai Group was first described by Dr. Edwin McKee for exposures on the Havasupai Indian Reservation in Grand Canyon. It is composed of four separate and distinct formations, three of which have been given Havasupai family surnames. These are, from oldest to youngest, the Watahomogi, Manakacha, and Wescogame, which are

© Wayne Ranney

The Supai Group (lower 2/3 of photo) in Carroll Canyon

all capped by the Esplanade Sandstone. It is difficult for most people to recognize the four individual formations because each is composed of repetitive layers of red sandstone and **mudstone** mixed with some gray limestone and **conglomerate**. Only an expert who closely examines these rocks can distinguish these four formations. They will be referred to hereafter simply as the Supai Group.

If the four formations of the Supai Group are so similar in appearance, why did Eddie McKee differentiate between them at all? Because he found unconformities within the layers, each of which documents an important period when non-deposition or erosion occurred. These unconformities provide important details for the story. It is known, for example, that 600 feet of the Supai Group accumulated over a period of 30 million years—on average one foot of sedimentation every 50,000 years. This is an extremely slow rate of deposition. However, the unconformities reveal that sediment was not accumulating continuously for all of these 30 million years; perhaps it was even being partially eroded at times. Keeping this in mind, it is likely that many deposits were laid down rapidly (possibly within a few months or years), but only a

UNCONFORMITIES

It might be difficult to imagine something that doesn't exist in the landscape, but as it turns out, the amount of time represented by rocks we can see today is just a fraction of all the time that has passed. Geologists have discovered that there are long periods of time that are not represented by strata in the rock record. These gaps are called unconformities. Since this concept might be difficult to comprehend at first, an example might be in order. Imagine that you are reading a book, and as you turn the page you notice that something just isn't right. Maybe the young boy you were reading about is suddenly portrayed as a middle-aged man. He lives with his wife and children now, whereas he had just been riding his bike with his boyhood friends. Of course, you would notice this discrepancy and your curiosity would cause you to turn back the page, where you discover that for some odd reason page 100 is followed directly by page 300. When you first picked up the book it looked normal, with page after page in simple succession. But in your "exploration" and "research," you discovered an "unconformity" of about 200 pages. As a reader, you knew something was amiss but it was only with the help of the page numbers that you had the first-hand evidence that pages were missing.

The "page numbers" that geologists use to verify the gaps between otherwise normal-looking strata are things like fossils or other materials in the rocks that can be easily dated. Good evolutionary sequences do exist for some organisms and show their slow transition from one form to another through time. If rocks contain fossils that show a disrupted sequence of change, then an unconformity is indicated. Adjacent layers may also contain datable materials such as volcanic rocks or ash, and these can be taken to the laboratory and shown to be of considerably different age, indicating an unconformity.

few survived the erosion episodes represented by the unconformities. These gaps in the rock record show that sedimentary layers undergo periods of rapid, active accumulation that are interspersed with periods of non-deposition or even erosion. This is how some beds can be deposited in only a few months' time, even if only 600 feet of sediment were preserved in 30 million years. In reality, most sedimentary rock units preserve only a fragment of all the geologic time that has passed.

During Supai deposition, the Sedona area was a low-lying coastal plain adjacent to a shallow sea. The **relief**, or elevation change, on this landscape was subdued and a slight change in sea level would cause the sea to inundate large portions of the coastal plain. The Supai, with its bands of alternating sandstone, mudstone, conglomerate, and limestone, records the numerous times that the seas washed over and retreated from the Sedona area. A good modern analog for this setting may be the Middle East where the low coastal deserts of the Gulf States and Kuwait straddle the shores of the Persian Gulf. If we could observe the Persian Gulf through time we would see it

Unconformities are quite common in the rock record and may represent periods of any duration. Some gaps in the Supai Group may represent only a few tens of thousands of years or they may be of one or two million years. The unconformity between the Tapeats Sandstone and the Martin Formation near Jerome, Arizona, represents almost 150 million years. And The Great Unconformity, between Jerome's ore-bearing rocks and the overlying Tapeats Sandstone, represents a whopping 1,200 million years— more than one-fourth of the entire history of the earth. This truly is a great unconformity.

An unconformity representing 265 million years is exposed on Wilson Mountain between basalt lava and white sandstone

Unconformities arise in two different ways, either by the initial deposition of sediment followed by its partial erosion, or by the complete absence of deposition for an entire period of time. It is often impossible to tell which of these two might be responsible for an unconformity, but geologists don't bother themselves too much about it. They just accept that a period of non-preservation is present in the sequence, and every bit of evidence found, even the void that signals an unconformity, helps them to unravel the myriad mysteries of Earth's history.

expand and contract over these low, coastal deserts, leaving stratified layers that record the sea's rise and fall. Now, as then, the mechanism causing these sea-level changes is the expansion and contraction of glacial ice sheets far away.

Towards the end of Supai deposition, sand **dunes** blew in from the north and covered the region. Evidence for this is preserved as the Esplanade Sandstone exposed in the small canyon below Midgley Bridge. The Esplanade is primarily composed of sandy dune deposits, with only a very few limestone beds found at its base. This indicates more terrestrial dune conditions during the Pennsylvanian and continuing into the Permian. This trend is evident in other layers on the Colorado Plateau as well and continues with the next red-rock unit in Sedona, the Hermit Formation.

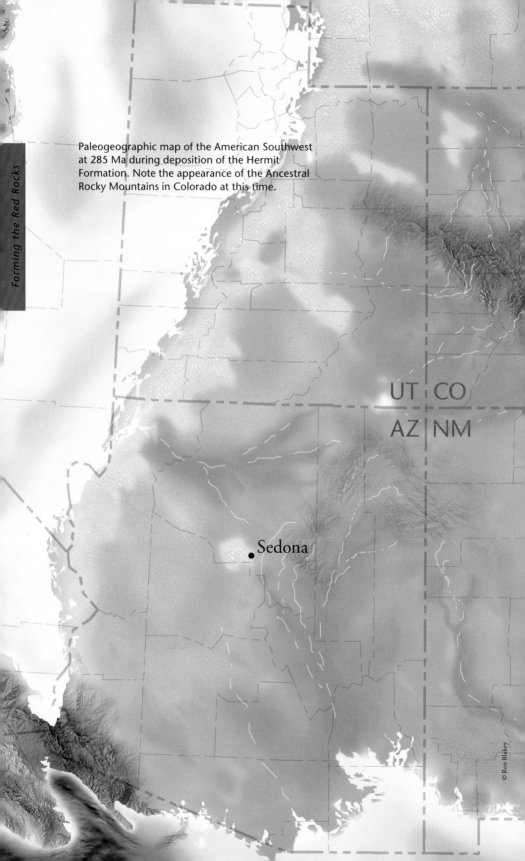

Paleogeographic map of the American Southwest at 285 Ma during deposition of the Hermit Formation. Note the appearance of the Ancestral Rocky Mountains in Colorado at this time.

UT CO

AZ NM

• Sedona

The Hermit Formation:
OF RIVERS AND
DISTANT MOUNTAINS
285 Million Years Ago

The Hermit Formation is something that Sedona residents interact with on a regular basis whether they know it or not. A large portion of the city is built upon it. **Outcrops** of the Hermit tend to form an extensive bench or a wide terrace that overlies the harder Supai Group rocks. This terrace is ideal for development – more so than the steep red-rock cliffs that rise above it. West Sedona and the Village of Oak Creek are built upon such an eroded terrace. Because of the way the Hermit Formation weathers it is often covered with thick growths of pinyon and juniper trees. There are few places where one can actually see the Hermit Formation and ironically this makes it readily distinguishable from the more resistant cliff-forming units above and below it. When new construction takes place in Sedona, geologists flock to these fresh scars in order to document the secrets exposed, knowing that this window into the past may be open for only a month or two.

When the Hermit Formation is seen in roadcuts however, as along Highway 89A west of the Sedona post office, it reveals evidence for deposition in a **fluvial**, or river environment. It often contains mudstone beds that are cut into and inset with conglomerate lenses. The mudstone has been interpreted as a river floodplain deposits.

The broad pinyon-covered terrace that marks the Hermit Formation in West Sedona. Bear Mountain is visible on the skyline.

© Wayne Ranney

The floodplain of a river is the land that lies outside the river channel, which is inundated only when the river floods. For instance, the floodplain of the Mississippi River is from ten to twenty miles wide, while its actual channel is less than a mile wide.

The rivers coursing across northern Arizona during Hermit time were probably not as big as the Mississippi, nor did all of them contain water year-round. However, they behaved in a similar way by altering their course many times after floods. The evidence for this is seen in the way the conglomerate and mudstone is preserved in cut-and-fill channel features. Often the conglomerates are preserved in concave channel forms, whose walls are composed of mudstone. Sometimes it is the other way around with mudstone found in a channel cut into the conglomerate. Both situations record small rivulet channels that were cut into the underlying sediment and filled later with the other material. Higher energy floods left conglomerate, low energy floods deposited mudstone.

A conglomerate-filled channel in the Hermit Formation

© Wayne Ranney

A good place to view this evidence is in the parking lot of the U.S. Forest Service visitor center on Highway 179, south of the Village of Oak Creek. Here, in the roadcuts west of the parking area, small scale channels can be found that were filled alternatively by mudstone or conglomerate. Walk around this roadcut and have a look. You may notice some channel forms filled with gravel and others with mudstone or sandstone. These cut-and-fill features can also be seen in the parking lot at Midgley Bridge and in the roadcuts exposed to the west of the Sedona post office.

Geologists have identified two types of channels near Sedona—small ones just a few feet in dimension and broader ones that are over half a mile wide. The small channels' morphology (shape) suggests that these may have been similar to modern southwestern **arroyos** where runoff occurs only after heavy rains. The rest of the time the arroyo is dry. The cut-and-fill channels are often covered by flat-bedded mudstone, indicating that the rivers once again changed their course as the area was buried under a wide mantle of floodplain deposits. One could look for days at these interesting relationships throughout Sedona and be impressed with what is preserved.

The Hermit Formation also contains many silver dollar-sized nodules of calcium and limestone similar to those found in the desert soils near Phoenix and Tucson today. These **caliche** deposits form when high daytime temperatures literally pull groundwater towards the surface. When the water evaporates a hard crust of minerals remains in the upper few feet of soil. The nodules of calcium in these rocks are evidence that the region was the site of arid river floodplains at this time. A few fossils can be found in these deposits, which further support this idea. One known as Walchia is an extinct plant that is thought to have flourished in arid environments. **Root casts** can also be seen in many outcrops, showing where the roots of these or other plants gained a foothold on the river plain.

If your interest is piqued by the story of the Hermit Formation, you've probably already asked where those ancient river systems may have originated. Sleuthing through the modern landscape can provide the answer. Deposits of the same age as the Hermit Formation have been preserved and exposed in other locations across the Colorado Plateau. These deposits have been correlated through fossils and sediment types that reveal in great detail what the geography might have looked like during this time. How is this done? By careful study of minute details in these other rock layers and comparing them to what is found in the rocks near Sedona. Let's examine just a few of these details.

When geologists study a particular rock stratum they look at how its interior **texture** is arranged and distributed. Geologists want to know if the **clast**s (grains) in a sedimentary rock are coarse-grained (pebbles that form conglomerate), medium-grained (sand that yields sandstone), or fine-grained (silt or clay that forms mudstone or shale). They look to see if the clasts are well-rounded (smooth) or poorly-rounded

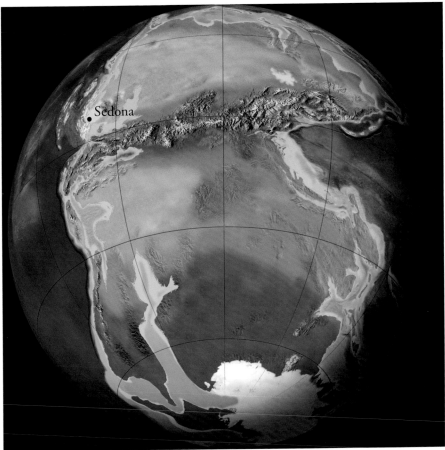

Global view of the earth's continents about 285 Ma during deposition of the Hermit Formation

© Ron Blakey

© Wayne Ranney

The Cutler Group exposed near Moab, Utah, is time equivalent to the Hermit Formation near Sedona. Note the pebble-size clasts in this rock indicating a source area in the Ancestral Rocky Mountains.

(angular). This yields information about how far the grains might have traveled. They also look to see if the rocks contain just one type of clast, such as quartz, or if they contain many types, such as quartz, mica, feldspar, and other minerals. A deposit with a variety of grains is termed immature because the feldspars and mica usually cannot withstand long transport in a river. Sediment with only quartz is termed **mature** because all of the other mineralogies have been winnowed out by long physical transport and chemical weathering. Another consideration is whether the clasts are well-sorted (of the same size) or poorly-sorted (containing various sizes). Finally, what kind of bedding is present in the stratum? Is it flat-bedded or perhaps not bedded at all? All of these aspects of a mudstone, sandstone, or conglomerate are important in determining the source area of the sediment, what kind of rock was being eroded, and in what kind of environment the sediment ultimately was deposited.

Regarding the Hermit Formation, geologists have noted that it and its related deposits change character from southwestern Colorado through the Canyonlands and Monument Valley areas, southward towards Sedona. What they see is that the clasts become generally finer-grained, more well-rounded and well-sorted, and more mature in a southwesterly direction. This **trend** in the maturity of the deposits indicates a source area that must have been located somewhere in southwestern Colorado. The evidence shows that a mountain range must have existed in southwestern Colorado along a line where the cities of Green River, Utah, Grand Junction and Durango, Colorado, and Taos, New Mexico, are now located. To the northeast of this line, no rocks of Hermit age or older are preserved. Therefore, this must be the location of an ancient range of mountains, which geologists have named the **Ancestral Rocky Mountains**. The Ancestral Rockies have no connection whatsoever with the present-day Rockies; the name simply denotes their former location where the modern Rockies exist today. No one ever saw the Ancestral Rockies—they developed millions of years before humans even evolved. Yet we know they were there because we are able to recognize the debris that washed out of them.

Amazingly, this ancient debris even tells us what kind of rock could be found in the Ancestral Rocky Mountains. The Cutler Group is a formation in Utah that is correlated with the Hermit Formation and it contains sandstone and conglomerate with chunks of angular granite. These informative exposures (seen in the photo to the left) represent the alluvial fan deposits that were washed onto a plain at the very foot of the Ancestral Rockies. The Hermit Formation near Sedona represents the near-coastal deposits from these rivers, although it is likely that many of the small channel deposits near Sedona were derived from rivers that developed locally, much like rivers today that have developed on the Great Plains below the modern Rockies.

This is the true magic of geologic study. After one has completed the tedious work of measuring the clast size within a sandstone or documenting the presence of calcium nodules in a mudstone, they can use the same tedious measurements made by dozens of other geologists in separate areas, and create a picture of how the region looked millions of years ago. This is how ancient landscapes are resurrected from the past.

© Bronze Black

Panorama towards Wilson Mountain

WHY ARE SEDONA'S ROCKS RED?

Perhaps the question most often asked about Sedona's landscape is, "Why are the rocks red?" To people from areas outside the southwestern U.S. this may seem to be a strange color for earth. Out here however, it's the norm; many areas on the Colorado Plateau have a lot of red rocks exposed. Said simply, the red rocks are the result of a thin layer of iron oxide that coats the outside of each and every individual grain in the sandstone. The iron giving Sedona's red sandstones their color represents only about one-half of one percent of the rock's weight. Though an individual sand grain from within the red rocks would barely look pink, very little iron is needed to create the deep red colors of Sedona's rocks. Iron could never be mined from here.

© Bronze Black

Sedona's sandstone was deposited in terrestrial environments—on river floodplains or wind-blown dunes. The source areas for the sediment were ancient mountain ranges that contained crystalline rocks like granite and schist. These types of rocks are composed dominantly of three mineral types—quartz, feldspar, and iron-rich minerals such as mica, hornblende, or pyroxene. Quartz is the most durable of the three and survives long transport distances as sand-size grains. Feldspars weather faster, turning into clay-size particles, which may come to rest in mudstone or shale. The iron-rich minerals (such as mica and hornblende) survive weathering and transport distances somewhere in between these two and can be deposited as grains along with the white quartz that makes up the bulk of a sand or mud deposit.

After burial of the sediment, groundwater flushes through it and attacks the iron minerals chemically. The isolated iron-rich grains act like little packets of red dye that can coat the white sand grains in the groundwater environment. Groundwater then carries iron-rich solutions "downstream" into layers of clear, white sand. Over time, these solutions leave a red coating on each individual sand grain and in this way the rocks turn red while still buried. Erosion eventually exposes the colored rocks.

Occasionally, thin white horizons can be seen within the red strata. These, too, are composed of sandstone but the sand grains at these horizons are just a bit larger than other layers. Because the grains are larger, there is more open space between the grains and this porosity allowed groundwater to move more readily through these horizons when it was all still buried. This increase in groundwater flow flushed away the red coating that used to line the grains of sand here. Look for these layers as you explore Sedona. (See the sidebar on page 49.)

The Merry-Go-Round on Schnebly Hill Road

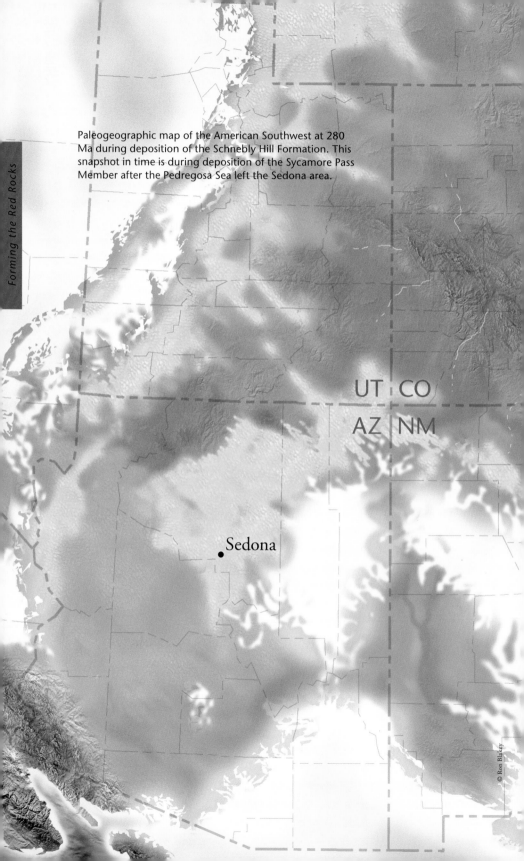

Paleogeographic map of the American Southwest at 280 Ma during deposition of the Schnebly Hill Formation. This snapshot in time is during deposition of the Sycamore Pass Member after the Pedregosa Sea left the Sedona area.

UT CO
AZ NM

Sedona

The Schnebly Hill Formation:
COASTAL SAND DUNES
280 Million Years Ago

The Schnebly Hill Formation is responsible for most of the spectacular red-rock scenery in the Sedona area. Bell Rock, Courthouse Butte, Cathedral Rock, and Coffee Pot Rock are all carved from this orange sandstone and all of the bright red rocks in the middle parts of Oak Creek Canyon belong to the Schnebly Hill Formation. These rocks are readily distinguished from other rocks in the area because they form tall, resistant spires that typically are a light orange color, compared to the darker salmon-colored rocks of the Supai or Hermit formations. The

Exposures of the Schnebly Hill Formation

Schnebly Hill Formation reveals a fantastic record of the environmental changes that occurred here approximately 280 million years ago.

Almost all of the Schnebly Hill Formation is composed of mature, quartz sandstone or siltstone. These rocks generally record a time when large, sandy dune fields existed in the Sedona area, much like the present-day Namib Desert in southwest Africa. Evidence from within the Schnebly Hill Formation shows that these dunes migrated from

Cross-bedding in the Schnebly Hill Formation

northwest to southeast. This evidence comes in the form of **cross-bedding**, which refers to the sloping layers of sandy strata found in otherwise flat-lying formations. Cross-bedding is always preserved on the lee side of a sand dune and thus records the direction in which the dunes were migrating. If cross-beds dip to the south, then the average direction of transport must be in that direction. In northern Arizona, most of the cross-beds in wind-derived sandstones dip to the south or southeast, recording a time when the dominant wind direction was from the north. Look for cross-bedding in the rocks while you explore Sedona.

A small unconformity, representing perhaps just a few million years, separates the Schnebly Hill Formation from the underlying Hermit Formation. This unconformity may be the result of the gradual erosion of the Ancestral Rocky Mountains, which would have reduced the supply of sediment available to the area. Perhaps there was a shift in the region's climate towards more arid conditions. This too could have diminished the sediment supply. Or perhaps

Courthouse Butte

© Wayne Ranney

the earth's crust in the Sedona area was simply not subsiding at this time and sedimentation was precluded by the lack of **accommodation space**. It is possible that all three of these played a role in this small unconformity.

The Schnebly Hill Formation is divided into four subdivisions or **members** in the Sedona area. Overall, a formation represents deposition in a specific environment, but occasionally a formation will show subtle changes within its strata, and these differences may be formally named as members of a formation. These most often record minor but important environmental fluctuations during the period of deposition. In the Schnebly Hill Formation, an **eolian**, or wind-dominated, environment is generally preserved, but variations in this setting are preserved within the four members.

Rancho Rojo Member – The lowest sandstones of the Schnebly Hill Formation are bright orange in color and are called the Rancho Rojo Member, after the subdivision of the same name in the Village of Oak Creek. This sandstone is only twenty to forty feet thick and is found only in outcrops near the Village of Oak Creek or along Wet Beaver Creek to the east. Because of its striking orange color the Rancho Rojo Member is easily distinguished from units above or below it. This is one of the relatively rare instances where the color of a rock unit can be used to determine important distinctions in the rocks such as formation boundaries or the recognition of faulted offsets.

The Rancho Rojo Member formed in eolian dune settings that were adjacent to nearby shallow-water environments. The evidence for this is found in the low-angle cross-beds that contain **adhesion ripples** on their upper surfaces. These form when dry sand is blown over wet sand and sticks or adheres to the wet sand. As the sand grains progressively pile up next to one another the adhesion ripples migrate upwind. The wet sand was found near the water's edge and as the tide went out, dry sand may have blown onto it forming the adhesion marks on the small eolian dunes. The Rancho Rojo Member originated as very low-lying dunes that were occasionally affected by the **nearshore** environment.

Bell Rock Member – In sharp contrast with the Rancho Rojo Member is the Bell Rock Member of the Schnebly Hill Formation. The Bell Rock Member is over 500 feet thick near Bell Rock, and constitutes the greater portion of this prominent landmark. This slightly darker sandstone is composed entirely of horizontally-bedded sandstone and siltstone. This causes the Bell Rock Member to erode into sweeping terraces of rounded, **slickrock** terrain, providing outdoor enthusiasts with one of the most pleasant walking environments in the area. It may be a comfortable coincidence that most of the so-called vortexes in the Sedona area are located on exposures of the Bell Rock Member; these weathering characteristics make for pleasant perches from which to view far-reaching vistas.

Ripple marks are sometimes preserved on the surface of the flat-bedded siltstone and these have a symmetrical shape on either side of their crest. This documents that the currents that formed them came from two or more directions. The horizontally-laminated sandstone and siltstone, and the symmetrical ripple marks suggest that the Bell Rock Member was deposited in a broad tidal zone where sediment was actively reworked by the tides. This explains why Bell Rock displays an

Ripple marks along the trail to Cathedral Rock

© Wayne Ranney

obvious lack of cross-bedding and a weathering profile that reflects its flat-bedded nature. But where did the sand originate? What brought it to the tidal zone?

Close inspection of the individual sand grains reveals the answer. The grains have "frosted" surfaces when viewed through a microscope. This results from the way light is refracted from behind small cracks and chips in the grains. These imperfections are thought to form when the grains are jostled about in an agitated environment. Water acts as a cushioning agent against the formation of these cracks, so it is unlikely that the sand originated in rivers or the tidal zone. Wind however, can cause the sand to become chipped. The sediment within the Bell Rock Member most likely arrived in the Sedona area on the wind, but was ultimately deposited and preserved in a broad, tidal zone of an ancient sea.

Fort Apache Member – An observant person who looks closely at the capstone on Bell Rock will notice a flat, gray deposit that contrasts sharply with the rounded red beds of the underlying Bell Rock Member. Previously called the Fort Apache

Limestone, this limestone, **dolomite**, and siltstone deposit is now called the Fort Apache Member of the Schnebly Hill Formation. In the Sedona area, the Fort Apache Member is only about ten or twelve feet thick but this actually makes it one of the more important units. In a

Wonderful exposure of the Fort Apache Member near Schnebly Hill Road

© Wayne Ranney

land rich with red sandstone, it is a blessing to have an extensive, thin horizon of gray rock to help locate faults and determine the amount of offset along them.

If we trace the Fort Apache Member to the west of Sedona, it pinches out entirely in the red sandstone cliffs in Boynton Canyon. To the southeast, towards its **type section** near old Fort Apache (at Whiteriver, Arizona), it is more than one hundred feet thick. This greater thickness to the southeast indicates that the depositional basin was centered in that direction. The thinning to the northwest records a short-lived but progressive **onlap** of the sea into the dune field environment of the Schnebly Hill Formation. This sea made one gallant advance as far as Boynton Canyon, evidenced by the pinching out of the gray rocks in that area. The unit is thicker to the southeast of Sedona, therefore the sea persisted longer in that direction. We have already noted that the Bell Rock Member is sand reworked by the tides. Now we can say with some certainty where those tides originated. The Fort Apache Member documents the incursion of the Pedregosa Sea into the Sedona area at this time.

Sycamore Pass Member – The uppermost unit of the Schnebly Hill Formation, the Sycamore Pass Member (named after exposures near Sycamore Canyon), demonstrates that desert dunes encroached over the Sedona area after the Pedregosa Sea withdrew to the southeast. It is composed of red-to-orange, high-angle, cross-bedded sandstone, indicating deposition in eolian settings without any reworking by the tides. At Sycamore Pass much of the 700-foot thickness of the Schnebly Hill Formation is composed of this member, with very little Bell Rock Member and no Rancho Rojo or Fort Apache members present at all. East of here, towards Sedona, the Sycamore Pass

WHITE STRIPES IN THE SCHNEBLY HILL FORMATION

People often notice the horizontal white stripes that can be seen among the red layers of the Schnebly Hill Formation and wonder what they are. They may even mistake these for the Fort Apache Member of that formation. They are not. There is only one Fort Apache Member and it is a muted gray color rather than the white of these other narrower stripes. So what are these and what do they represent?

Close observation shows that these white horizons are composed of sand grains that are somewhat larger in size than those in the horizons above and below them. Because the grains are bigger, there is more empty space between the grains, thus allowing groundwater to flow more readily through these horizons. This increased flow of water ultimately leached the red coating from the sand grains, giving these zones their white color. The layers above and below these stripes are more tightly packed with smaller grains and thus did not experience as much groundwater leaching through time.

Why are there larger grain sizes in some horizons? It could be that for short periods of time the wind was much stronger and able to carry larger material into the depositional system. The finer material was blowing around as well at these times, but would have blown farther away in the fierce winds. Near the shoreline of the Pedregosa Sea the winds were steady for most of the time and deposited fine- to medium-grained sand in most instances. But short bursts of much stronger winds brought more coarse sand into the system. It's amazing to think that such short-term ancient events can still be detected in today's landscape.

© Wayne Ranney

White horizons in the Schnebly Hill Formation

CORRELATION OF ROCK STRATA

Layers of strata found in Sedona can also be seen in other places on the Colorado Plateau, even though they might look different. For example, Sedona's rock formations (with one notable exception) comprise the upper one-third of the Grand Canyon's walls. From the viewpoint on Airport Mesa, look north through Soldier Canyon and you'll see the close resemblance between the Mogollon Rim here and the North Rim of Grand Canyon. But "good looks" get you only so far in geology. Rocks that look alike must also be shown to have a temporal connection to one another before their relationship can be confirmed, and this is where the art of correlation is used.

It is just a little more than eighty miles "as the red-tailed hawk flies" between Sedona and the Grand Canyon. The strata appear to be quite similar and continuous across this distance even though they are completely buried between the two areas. Recognizing the same strata to the east, however, becomes problematic because the strata in that direction are either still buried (and are known only from well-drilling logs) or they have changed their facies (character) when they do appear again from the subsurface. Someone untrained in the art of the correlation would not be able to tell that these rocks were related in time to Sedona's red rocks.

To correlate rock layers it is important to know that strata are deposited in an observable sequence that is never "shuffled." For example, in Sedona or anywhere else it is exposed, the Supai Group is always found below the Hermit Formation and never on top of it. The sequence of layers is always the same even if some layers are missing. Rock units might pinch out resulting in a sequence with fewer layers. Or one rock type might gradually change into another and the two will look nothing alike. In these situations more detailed studies must be carried out in order to determine how the rocks relate to one another.

Geologists use a variety of techniques to determine how the sedimentary layers correlate to one another. One good technique employs index fossils—fossils that are widespread in extent but restricted in

© Wayne Ranney

Coffee Pot Rock frames a view of the Mogollon Rim

time. If two rocks look different but contain the same index fossil, they must be of the same age. If the sediment contains thin layers of volcanic ash, the ash can be dated to give an age. Volcanoes throw ash far and wide and this material can be dated by radiometric methods with accuracy. Perhaps the best correlation technique is simply to follow an outcrop along its horizon to see how this horizon "plays out" across the landscape. This tool is especially useful in the American Southwest where the exposures are excellent and often uninterrupted. When rock layers go underground or disappear, they become more difficult to correlate, but knowing the sequence of strata usually can be of great help.

Member begins to interfinger with and grade laterally with the Bell Rock Member. The Fort Apache Member begins to separate the two in Boynton Canyon. Most of the Sycamore Pass Member is missing in the southeast near Fossil Creek (twenty miles southeast of Sedona). These relationships provide us with a bird's-eye view of the ancient landscape about 280 Ma.

Looking at the big picture, the four members of the Schnebly Hill Formation tell us that marine environments were more prevalent southeast of Sedona, while eolian dune conditions dominated to the northwest. This trend is clearly evident on a small scale in the Fort Apache Member: it pinches out in Boynton Canyon, is ten to twelve feet thick near Sedona, and more than one hundred feet thick at Fort Apache. Thus we know what relationships existed between desert dune and marine environments in a lateral or spatial dimension.

We can also determine the vertical, or time dimension, by looking at the inter-tonguing relationship between the Sycamore Pass and Bell Rock members. Remember, the upper layers of rock are younger than those they cover. As noted, the highly cross-bedded Sycamore Pass Member makes up most of the 700 feet of the Schnebly Hill Formation in the west near Sycamore Canyon. For most of Schnebly Hill time, desert dunes dominated the landscape there. Towards the southeast near Sedona, the Sycamore Pass Member gradually interfingers with the Bell Rock Member. And farther southeast still, at Fossil Creek, the Sycamore Pass Member is gone entirely and the Bell Rock Member dominates. This tells us that near-shore conditions existed in the Sedona area early in Schnebly Hill time, but that desert dune environments became dominant in later times. It shows that the dunes (Sycamore Pass Member) gradually overwhelmed the earlier near-shore environments (Bell Rock and Fort Apache members) through time.

Time-equivalent deposits reveal what was happening in other parts of the Colorado Plateau. The thickest deposits are located in the subsurface beneath the town of Holbrook, Arizona. (Cores from water wells tell us what's there.) More than 2,000 feet of Schnebly Hill Formation are present in the subsurface beneath Holbrook. This is almost three times the thickness of that preserved near Sedona. This thick layer indicates a rapidly subsiding basin that accumulated a great deal of sediment. Equally impressive is the dramatic transformation the sediments have undergone. The area around Holbrook was located closer to the source of seawater during this time, and most of the unit is composed of salt layers laid down in evaporative salt flats.

The Schnebly Hill Formation was also deposited near present-day Williams and Cameron, Arizona, but is present only in the subsurface. These layers come to the surface however, to the east in Monument Valley and in Canyon de Chelly, hence the name De Chelly Sandstone used there. No Schnebly Hill Formation is preserved in the Grand Canyon, suggesting that this area was either too high for sediment preservation or that the dune field did not extend that far north. Sedona thus has a rock formation that its more famous cousin does not.

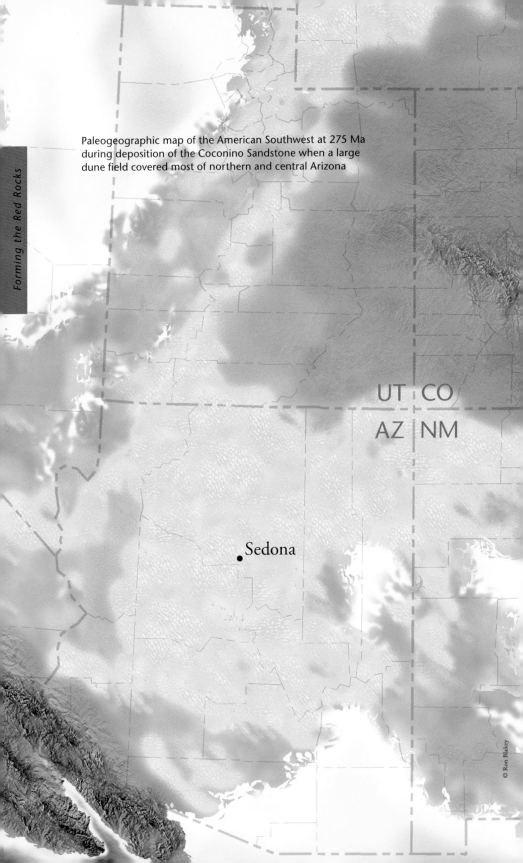

Paleogeographic map of the American Southwest at 275 Ma during deposition of the Coconino Sandstone when a large dune field covered most of northern and central Arizona

UT CO
AZ NM

•Sedona

The Coconino Sandstone:
INLAND SAND DUNES
275 Million Years Ago

As the Pedregosa Sea withdrew to the southeast and away from the Sedona area, the sand dunes migrated in step with the shoreline in that same direction. Coastal dunes gradually became inland dunes in the Sedona area. There was no clear break in deposition between the coastal and inland dune periods, and a **gradational contact** was developed. This type of **contact** is exposed today as a 100-foot-thick

Gradational contact between the Schnebly Hill Formation and the Coconino Sandstone

sequence of red Schnebly Hill Formation interbedded with white or colorless Coconino Sandstone. It is impossible to put a finger on the exact horizon where one formation turns into the other. (Those who need to define a boundary usually use the highest occurring red sandstone as a marker.) Gradational contacts are relatively rare in the geologic record because environmental change usually brings about periods of erosion or non-deposition in sedimentary basins, leaving us with an unconformity. But the gradational relationship between the Schnebly Hill and Coconino formations records unceasing deposition in this ancient sandy desert.

The Coconino Sandstone is named for exposures on the Coconino Plateau south of the Grand Canyon. It is preserved across an area from near Lake Mead, Nevada, to northern New Mexico, and from the Arizona-Utah state line down to central Arizona. This rather large tract of sandstone represents deposition in the heart of a great inland dune field, areas known as **ergs** (large sand "seas") in modern Middle Eastern deserts. These inland dunes are subtly but recognizably different from those in coastal dune environments. Inland dunes typically have uniform, larger grain sizes and cross-bedding throughout rather than flat- or trough-bedded sandstone. In fact, the larger size of the sand grains in the Coconino played a major part in making it

53

a white sandstone rather than red. White sandstone occurs because groundwater (or hydrocarbons in some instances) can move more readily through strata that has larger, uniform grains. Think of a pickup truck full of volleyballs. Is the truck truly filled with volleyballs or would it be possible to store some water in the voids between the rounded balls? Of course, the answer is Yes, and sand grains behave in much the same way. Therefore, it is conceivable that the Coconino Sandstone used to be stained red, but groundwater or oil moving easily through it leached out any iron staining that may have coated its clasts.

Under a microscope the individual sand grains of the Coconino reveal a "frosted" texture on their surface. This frosted appearance is created because of the chips and cracks present on the surface of the grains. Light passes through these fractures and is

© Bronze Black

The Coconino Sandstone in Oak Creek Canyon

refracted back to one's eyes. These cracks were formed by the unceasing collisions between grains as they blew in the wind. Frosted surfaces on sand grains are one of the unique indicators that geologists use to deduce that sandstones have been deposited in eolian settings.

Additional evidence for an eolian origin of the Coconino includes the high-angle cross-beds, the numerous trackways of reptiles found on some cross-beds, and unique raindrop impressions. The angle of repose for loose sand is about 33 degrees to horizontal. If sand is piled any steeper than this it avalanches downhill. The lithified cross-beds in the Coconino are between 29 and 31 degrees, the lower angle due to post-depositional compaction of the sediment. Sand can only accumulate at 33 degrees in **subaerial**, as opposed to **subaqueous** (underwater), environments. High-angle cross-beds are thus very good evidence for deposition in eolian settings.

The trackways were created when reptiles made their way up the lee side of a sand dune immediately after a light rain. Wonderful sets of ten or more reptile foot imprints are sometimes found in the Coconino and Schnebly Hill formations. Perhaps even more amazing are the raindrop impressions that are rarely seen as small pits with raised rims, usually about one or two inches apart. These small imprints record a light, desert rain some 275 million years ago. It is easy to imagine these types of desert rains since they are oftentimes encountered in today's Sonoran Desert.

The Coconino has a significant impact on the economy of northern Arizona as its uniform cross-beds are quarried and shipped all over the continent to be used as flagstone for patios and walkways. The economy of the little town of Ash Fork, northwest of Sedona, is almost entirely dependent on this geologic resource. Many of these quarries can be seen cut into the forested veneer while traveling on Interstate 40 or U.S. Highway 89 south of Ash Fork. After 275 million years of silent and immobile rest, the sand grains of the Coconino Sandstone are in motion again, traveling on flatbed trucks to the patios and walkways of America.

Reptile trackway near Sedona

Cross-beds in the Coconino Sandstone

55

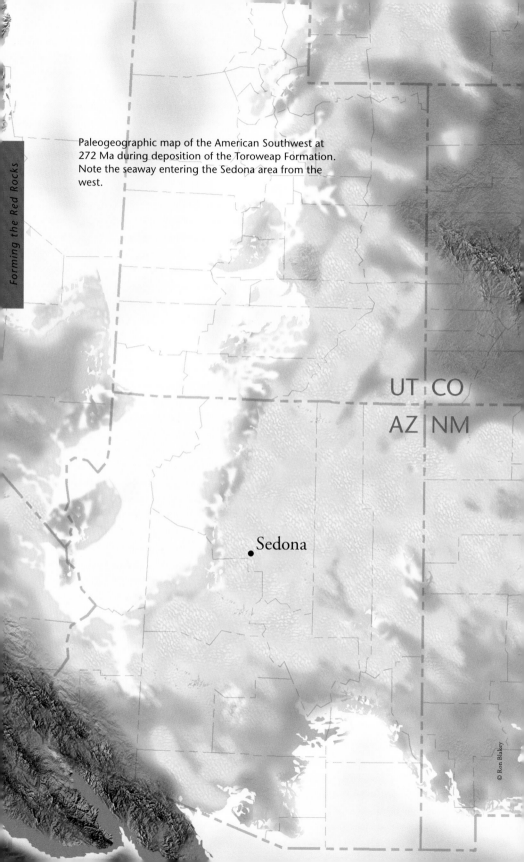

Paleogeographic map of the American Southwest at 272 Ma during deposition of the Toroweap Formation. Note the seaway entering the Sedona area from the west.

UT CO
AZ NM

Sedona

© Ron Blakey

The Toroweap Formation:
A CHANGING COASTLINE
272 Million Years Ago

Uninterrupted deposition continued beyond the end of Coconino Sandstone time but the environment evolved slowly nonetheless, causing the sediment type to change leaving the Toroweap Formation upon the landscape. This variable stratum yields an unequaled view into these changing landscape conditions. It also illustrates clearly how stratigraphic units can change, almost imperceptibly, in spatial (horizontal) and time (vertical) dimensions. Geologists, however, can perceive these very subtle variations (that's their job) and they provide us with a view of an evolving desert coastline some 272 Ma.

Any sedimentary rock unit extends for some distance in a certain direction but nowhere on Earth does any one formation continue forever and uninterrupted across the landscape. The earth is not covered continuously with any one type of rock. Many times rock formations become thinner and thinner until they pinch out. Sometimes they are **beveled** after deposition and covered by younger sediment. And occasionally a rock formation will gradually change from one rock type to another until it is so utterly different that it looks like another formation. This change of rock type may occur imperceptibly over many miles and may go unnoticed to the untrained

© Wayne Ranney

The coast of Namibia in southwest Africa is a perfect modern analog for the Toroweap Formation.

eye, or it may occur within the space of just a couple of miles in a transition that can be observed from a particular vantage point. The Toroweap Formation in the Sedona area undergoes a change somewhere in between these two extremes.

In a distance of only ten to fifteen miles between Sycamore and Oak Creek canyons, the Toroweap Formation undergoes a fantastic change from an inconspicuous, slope-forming siltstone, limestone, and **gypsum** deposit to a substantial cliff-forming, cross-bedded sandstone. This **facies** change (literally, change in aspect), documents with precision the difference in environments from one place to the next. In detecting this subtle change the ancient landscape can be envisioned as if observed from the vantage of a helicopter flying over a warm, shallow sea at one place but climbing onshore towards a wide, sandy beach. This ancient sea-to-shore transition may be observed in the walls of the Mogollon Rim north of Sedona.

The siltstone, limestone and gypsum facies of the Toroweap Formation was deposited in a shallow marine or **sabkha** environment. Sabkha is an Arabic word that describes coastal deserts whose soils are composed of salt or gypsum. There are many sabkhas in modern Middle Eastern deserts that are wide, flat, and treeless beneath the searing desert sky. In this setting the hot, near-ceaseless radiation from the sun draws groundwater through the soil and up onto the desert floor, a process known as capillary movement. When this saline groundwater reaches the desert floor it quickly evaporates and the minerals it carried with it are left behind in thin, crinkly crusts of salt and gypsum. These desert crusts can accumulate to considerable thickness through time. The area around Sycamore Canyon looked like this some 272 Ma.

Meanwhile, conditions were considerably different to the east, where Oak Creek Canyon is located today. Here the Toroweap horizon is a cliff-forming sandstone preserved with graceful cross-beds that can be seen in the roadcuts along Highway 89A near the West Fork. At first, it may be difficult to discern a difference between the Toroweap and the Coconino sandstones in this area. But there are subtle differences to be seen. The Coconino has steep, tabular cross-beds formed in a strictly eolian environment, while the Toroweap has curved, trough-shaped cross-beds that swirl and dip at various angles. These trough-shaped cross-beds were deposited by the coastal winds that came off of the sea to the west, while the siltstone, limestone, and gypsum facies was being deposited at the same time to the west. This explains why the Toroweap looks so different between Oak Creek and Sycamore canyons—the environments were vastly different.

Because the Toroweap is virtually indistinguishable from the Coconino in Oak Creek Canyon, one local geologist has proposed that the name Toroweap not be used there (see the discussion in Road Log 2 on page 130). Ron Blakey has proposed that all of the white, cross-bedded sandstone in the canyon be classified as Coconino Sandstone with two members. Under his scheme there is no Toroweap Formation in Oak Creek Canyon. This can be confusing to some people but the naming of rock units by geologists is an inexact science, especially when sediments undergo a relatively rapid facies change. Sometimes scientists attempt to draw lines where no such lines exist in nature.

I happen to be of the opinion that the use of the name Toroweap in Oak Creek Canyon can be justified for a few reasons. This is not to slight Dr. Blakey's idea; it's just that we each prefer to highlight different aspects of its outcrop characteristics. For example, it is possible to locate a sharp and definitive contact between the two formations, even in Oak Creek Canyon. This contact is known as the "green line" and it becomes obvious (when one becomes attuned to looking for it) in an otherwise un-interrupted cliff of white sandstone. The "green line" is formed because of a slight difference in the hardness of the two formations and a narrow ledge has formed on top of the Coconino, allowing trees and shrubs to gain a tenuous foothold. It can be seen from just about anywhere in Sedona wherever the contact is clearly exposed. Although the "green line" is an informal designation, it is easily identifiable. Dr. Blakey prefers to look at the composition of the rocks in making the case for calling the Toroweap an "upper" Coconino Sandstone; I prefer to accentuate the lateral equivalency of the Toroweap deposits. Even though we use different terminology we both agree upon what the rocks actually represent.

An excellent view of the "green line" in the Mogollon Rim west of Sedona

Another way to look at the relationship between the Coconino and Toroweap formations is to consider their combined thickness. Together these two units are about 800 feet thick and represent perhaps 4 million years of deposition. In Oak Creek Canyon this entire 4-million-year interval witnessed the deposition of sandstone, the lower part representing deposition as inland eolian dunes and the upper part as dunes much closer to the sea. However, in Sycamore Canyon the first 2 million years witnessed the accumulation of Coconino Sandstone, while the last 2 million years preserved siltstone, limestone, and gypsum deposited in a sabkha setting. When viewed in this way, one can more readily comprehend exactly how sedimentary rocks record the changing nature of our earth's landscapes.

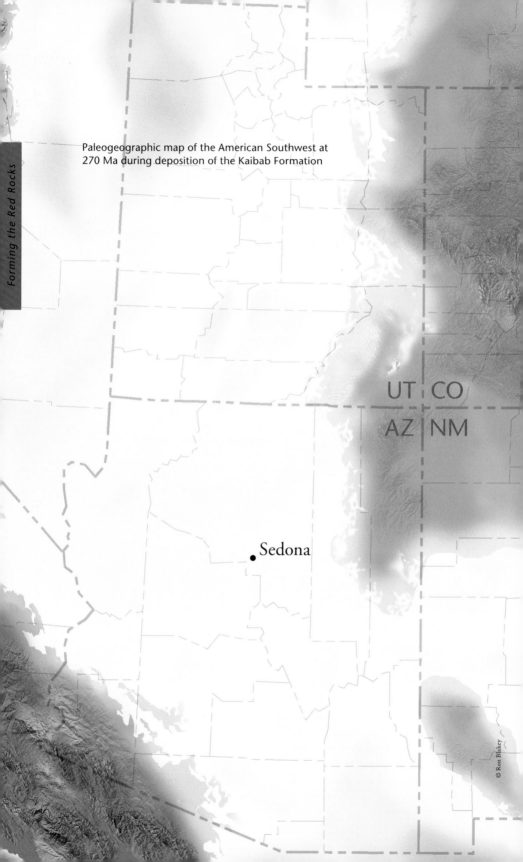

Paleogeographic map of the American Southwest at 270 Ma during deposition of the Kaibab Formation

UT | CO

AZ | NM

• Sedona

Forming the Red Rocks

© Ron Blakey

The Kaibab Formation:
ONE FINAL SEA
270 Million Years Ago

The layer of rock that caps the Mogollon Rim in Arizona is called the Kaibab Formation. This unit was formerly called the Kaibab Limestone but the name has been changed because of the many rock types found within it, namely dolomite (a different kind of limestone), **chert**, and minor sandstone. The Kaibab is approximately 350 feet thick in the Sedona area and is easily distinguished by its prominent horizontal bedding often forming overhanging ledges or steep slopes. In most places around Sedona it is the highest exposed rock, except when occasionally overlain by relatively recent, black volcanic rocks. The Kaibab is an important part of Sedona's geologic heritage because it is quite durable and resistant to erosion, which "holds up" and preserves all of the underlying strata.

To understand why the Kaibab is so resistant to erosion, it is helpful to become familiar with the conditions in which it was laid down. The Kaibab was deposited in the same sea that left the Toroweap Formation just a few million years earlier. But through time this sea advanced farther to the east, inundating all of northern Arizona with slightly deeper and clearer seawater that did not evaporate as readily. These marine conditions were more conducive to supporting sea life and allowed for more limestone, dolomite, and chert to be deposited. There was more open circulation of the seawater in Kaibab time, which allowed for a richer fauna that produced the limestone and chert. Great blooms of sponges grew in profusion on the sea floor, and when they died their spicules (silica needles) rained down on the ocean floor by the millions, accumulating and compacting into dense layers of white chert. This chert is found as oval-shaped nodules or thin beds within the limestone and dolomite of the Kaibab Formation. Deposition of the Kaibab Formation ceased about 270 Ma.

After this time, the Kaibab was subjected to about 30 million years of near-surface exposure. It is possible that other sediments were deposited on top of it during this time, but if so, they were eroded away before the next deposit was preserved. However, we know that groundwater percolated through the Kaibab Formation during this time because some of the limestone and gypsum in it were dissolved and replaced with chert. Over time the Kaibab became more enriched with chert, which is a harder and more resistant rock type. Were it not for these ancient blooms of sponges and the special conditions that existed after their burial, the Kaibab might have been a more ordinary, more easily dissolved limestone unit. This multitude of sponges and their billions of spicules made the Kaibab a resistant rock unit and thus saved the underlying formations from certain erosion and quicker removal. Without the resistant Kaibab Formation, the Mogollon Rim and certainly the Grand Canyon (also capped by chert-laden Kaibab) would not be the same spectacular landscape features that we see today.

Rocks and Wandering Continents:
RELATING SEDONA'S ROCKS TO OTHER ANCIENT LANDSCAPES

The degree of change that occurred in the Sedona area over a long span of 46 million years is quite impressive. Sedona witnessed incredibly diverse scenes in its colorful past: an equatorial sea, a river floodplain, windswept deserts, sandy coastlines, and one last tropical sea. Geologists have been able to resurrect these ancient landscapes that lay hidden in the red rocks for most of human history. Those lucky enough to become aware of this longer history of the scenery will forever see the land in a much different light.

Moving forward with our story, it's enticing to think how this region was connected in the past with other parts of the globe. In fact, the sedimentary rock record worldwide provides a stunning view of the global environment that existed between 316 and 270 Ma. It is now possible to place Sedona's red rocks within the context of a global framework and envision their relationship to amazing events occurring elsewhere on Earth. This chapter tells the story of how the red rocks near Sedona correspond and relate to other long-lost events that occurred on our planet.

When the Sedona area was an arid coastal floodplain along the western margin of North America (Supai Group), a momentous event began to unfold in a much different setting to the east. (It should be noted that directions are given with respect to modern coordinates. Because the continents move about or drift over the earth's surface, what appears east today may actually have been oriented in some other direction with respect to the earth's stationary axis.) For a few hundred million years prior to the Supai Group deposits, North America had been drifting slowly to the east. About 316 Ma it collided with the westward drifting continents of Africa and Europe. This mighty collision wrinkled the earth's crust and created a lofty mountain chain. The eroded remnants of this mountain range are today's Appalachian Mountains in North America and the Caledonian Highlands in Great Britain.

Ancient rivers carried great quantities of sediment out of these mountains and deposited it as alternating layers of sandstone, shale, and conglomerate in the lowlands that existed on either side of these mountains. The climate in these lowlands was humid, and dense forests of tropical vegetation grew beneath the towering mountains. The evidence of these lush forests on the west side of the ancient mountains is found in the great layers of coal that are mined in the states of Ohio, West Virginia, and Pennsylvania. Coal of the same age is mined in England, France, Germany, and Poland, indicating the presence of tropical forest on that side of the mountains as well. The quantity and quality of sedimentary preservation in these rocks are so good that geologists in the United States name this period for rocks found in one particular state where they are found in profusion—the Pennsylvanian.

This was a glorious and lively time on our planet. Dragonflies with three-foot wingspans have been found between layers of shale. The earth's first preserved reptile was found in rocks of this age, in an unbelievable setting—the hollow of a petrified log in Nova Scotia. Imagine the journey one could have made from the far-reaching coastal deserts of the Sedona area to the jungles at the foot of the ancient Appalachian Mountains, and then over their snow-capped summit to a similar forest on the eastern side. Our earth has witnessed such marvelous scenes.

When the Pennsylvanian Period ended and the Permian Period began 299 Ma, other continents drifted in from the south and amalgamated with the growing expanse of land. In fact, the collision of South America with North America is the most likely reason that the Ancestral Rocky Mountains were uplifted near the modern Four Corners region of the American Southwest. The roots of similar ancient mountains are found in a broad arc to the southeast of the Ancestral Rockies in the states of Texas, Oklahoma, and Arkansas. All of these mountains were formed in the same way as the modern Himalayas, where the northward drifting India subcontinent is colliding with the southern side of Asia. Both the modern Himalayas and the Ancestral Rockies lack evidence of any volcanic activity because when two continents collide there is no subduction of oceanic crust and thus no melting of rock to form volcanoes.

With the culmination of these events, all of the earth's continents were assembled into one giant landmass called **Pangaea**, Greek for "all land." The near-shore and shallow marine environments preserved as the Toroweap and Kaibab formations in the Sedona area represent deposition on the far western fringe of the Pangaean supercontinent. During this time one could have journeyed across all of the earth's continents, including Antarctica, without crossing a body of water larger than a river. Imagine starting from the far western edge of Pangaea, replete with its desert dunes and arid coastlines near Sedona, and traversing numerous mountain ranges thrust upward by the collisions of vast continents. Think of the many different life zones one would have seen and the strange animals and vegetation that have long since vanished from our planet. How amazing that our study of geologic strata allows us to recall this vanished world from so long ago.

Using this geologic evidence, we can position ourselves in an imaginary spacecraft flying high above the future red-rock country. With plenty of fuel, we can watch the landscape evolve through 46 million years of Earth history. At first, the view out our window shows the muddy deltas, sluggish rivers, and low-lying dune fields of the Supai Group as it buries the tropical marine deposits of the Redwall Limestone. To our right, we see South America approaching from the south and colliding with North America. The Ancestral Rockies are thrust up to the northeast of the Sedona area and a river system flows out of them delivering the pebbles, sand, and mud of the Hermit Formation. Through time, these river systems dry up and are choked off as wind-driven sand is carried down from the north along the west coast of Pangaea. This sand eventually comes to rest along the shores of the Pedregosa Sea, formed to the south of Sedona by the progressive deformation of the earth's crust in collision with South America. This leaves the colorful Schnebly Hill Formation in the Sedona area. As the Pedregosa Sea retreats to the south, the Coconino Sandstone marches in across the landscape. Soon the western sea laps onto the edge of Pangaea, at first driving the dune fields to the east and leaving the Toroweap Formation. However, as this sea continues to advance, the Kaibab Formation buries the remains of all previous environments. Through 46 million years, the winds and seas of time bring sweeping changes to the landscape.

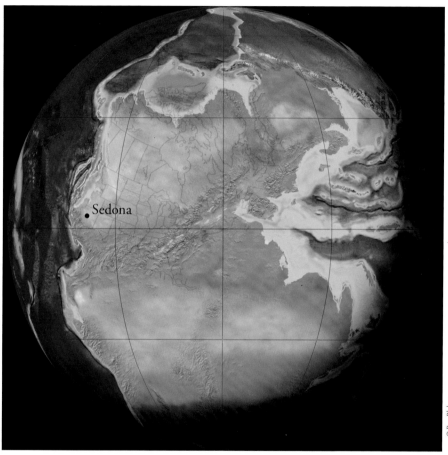

© Ron Blakey

A view of Pangaea with North and South America, Africa, and western Europe welded together along the Appalachian Mountains

Our ability to recognize the evidence for these changes is the gift of science and geology. The advances in geologic understanding, begun just a little more than two hundred years ago and accelerated in the last forty to fifty years, are nothing short of phenomenal. These discoveries and the far-reaching views they provide to us, appeal to our philosophical and spiritual selves as much as our inquisitive and scientific selves. As students of the earth, we have found the closest thing yet to time travel.

MESOZOIC INTERLUDE   250 TO 65 MILLION YEARS AGO

There is a gap in the rock record in the Sedona area that represents more than 225 million years of Earth history. And yet it is possible to reconstruct some of this history by examining the strata found in nearby areas on the Colorado Plateau and in the Basin and Range. Most of this time belongs to the Mesozoic Era, often referred to as the Age of Dinosaurs since they are the most visible life form from this era. It is likely that Mesozoic-age sediments did accumulate in the Sedona area but erosion has already removed them from the landscape. Throughout the rest of the Colorado Plateau a cornucopia of evidence remains, giving testament to what might have happened here during this time.

The colorful strata observed today in the Painted Desert of eastern Arizona or the walls of the Echo and Vermilion cliffs to the north were quite likely once present near Sedona. Perhaps even more deposits rested on top of these. In fact, on the Navajo and Hopi Indian reservations more than 10,000 feet of Mesozoic-age sedimentary rocks are exposed, and although it is impossible to know how much of this used to exist in the Sedona area, most likely a substantial amount did. These sedimentary rocks document the presence of many different landscapes including: desert rivers that meandered from a volcanic highland (the Chinle Formation), sandy deserts (the Wingate, Navajo, and Entrada sandstones), and long white beaches, lush swamps, and shallow seas (the Mesa Verde Group). It is likely that many of these changing scenes once graced the area around Sedona, complete with a large reptilian fauna.

© Wayne Ranney

Equally enticing to imagine are the moments in time when these sediments were eroding into the fantastic landscapes that must have existed here. Vistas like those observed today in the Painted Desert, Glen Canyon, or Zion National Park may once have been found in the Sedona area. Of course, the details of the landscape's actual appearance are lost forever. The types of vegetation that grew, the now-extinct animals that grazed the land (or each other), and the exposed rock strata are forever lost to even the most astute geology detectives. But we can say with confidence that these landscapes and these events occurred. We can look at the blue sky above Sedona and see much more than the passing clouds. One can feel the pull of an ancient river, or watch a dinosaur stalk the banks of a tropical swamp, or hear the waves crashing on a long, lonely beach. In the blue sky above Sedona, one can sense the ceaseless passage of time, the only constant on our wondrous, living planet. The Mesozoic Era, although not represented in Sedona today, certainly left its mark prior to our probing in the cliffs and canyons of the Mogollon Rim.

Oak Creek in the red rock country

PART 3

Sculpting the

Red Rocks

© Bronze Black

Paleogeography about 65 Ma when rivers flowed northeast from the Mogollon Highlands across Sedona

UT CO
AZ NM

. Sedona

Uplift and Erosion:
THE MOGOLLON HIGHLANDS
80 To 40 Million Years Ago

All of the geologic events discussed thus far occurred in landscapes vastly different from what is present near Sedona today. If a person were to be magically transported back in time to the top of a Coconino sand dune or a beach along the Kaibab sea, there would be no indication that these environments would eventually come to resemble the future red-rock country. All of this ancient history occurred very near sea level. But by about 80 Ma geologic forces began to lift and shape the land. These events signal a stupendous change from the long period of subsidence and accumulation of sediment, to one of massive uplift and erosion. Instead of the accumulation and burial of layer after layer, the colorful sediment would now be sculpted. Geologists use a fragmented or, in some cases, totally obliterated body of evidence to understand this part of the region's history and not all of them agree about what this evidence might be telling us. By using sophisticated techniques and a bit of detective fieldwork, however, we can discover much about how the red-rock country came to look the way it does.

At various times during the Mesozoic Era, central Arizona was uplifted into a mountain range that geologists have named the **Mogollon Highlands**. The name was given to this ancient range because of its proximity to the modern Mogollon RimA but the two features are actually separate and distinct landforms. Like the Ancestral Rockies before them, the Mogollon Highlands had completely eroded away before any human saw them. They can be detected indirectly, however, from three compelling lines of evidence: the erosion of most of the colorful sedimentary rocks from central Arizona, a slight northeastern tilt found in the red rocks near Sedona, and scattered bits of gravel that originated from the rivers that must have drained the highlands. Most of this evidence is not readily apparent to the casual observer but the events that created it had a profound effect on shaping the area's contemporary landscape.

The removal of sedimentary rocks from central Arizona and Mesozoic rocks in Sedona resulted from the uplift of the Mogollon Highlands. Geologists believe that the highlands formed when North America collided with pieces of Pacific Ocean crust far to the southwest of Sedona. This collision was initiated when Pangaea broke up about 180 Ma, causing North America to drift away from Africa and Europe. As North America moved west, (at about the rate that fingernails grow, or two inches per year) its southwestern edge collided with this ocean crust, crumpling and folding the edge of the continent, much like a throw rug being pushed along a hardwood floor into a wall. This collision created the strong compressive forces that uplifted the earth's crust into the Mogollon Highlands. It is important to note that central Arizona was uplifted higher in elevation relative to the Colorado Plateau and Sedona areas as a direct result of the collision. This event had huge implications in the formation for the modern landscape.

About 80 Ma, the Mogollon Highlands were located along a trend where the cities of Tucson, Phoenix, Kingman, and Las Vegas are located today. In the center of the highlands, sedimentary rocks similar to those still found in Sedona were intensely compressed, folded, uplifted, and subsequently eroded away. On the northeast side of the highlands, away from the most severe effects of the collision, the deformation of rocks was much less intense. Here the strata were only gradually arched up into a broad, tilted plain. These once flat-lying rocks developed a regional slope that dipped to the northeast at a nearly imperceptible angle of about 2 degrees. Geologists presume that as one would have approached the highlands from the Sedona area, the dip of the strata would become progressively steeper until very deformed and contorted rocks appeared on the crest of the highlands (see the diagram on page 81).

When the Mogollon Highlands formed, Mesozoic-age sediments in central Arizona were the first to be subjected to erosion. Through time these strata were stripped back to the northeast towards Sedona. For some time the colorful cliffs of this Mesozoic rock must have been exposed in the Sedona area but erosion eventually stripped them completely away. This retreat of the Mesozoic layers is still in progress and can be seen northeast of Flagstaff as a line of cliffs that make up the colorful landscape between Cameron and Holbrook.

As the Mesozoic rocks were being removed from the future red-rock country, erosion continued to attack older rocks in the Mogollon Highlands. This ultimately exposed the Pennsylvanian and Permian strata that once existed there, although it is impossible to know if these were precisely like the rocks present around Sedona today; they are completely gone now. When all of the sedimentary rocks had been removed from the highlands, a core of **Precambrian**-age crystalline rocks was **unroofed**. For the first time in almost half a billion years this ancient schist and granite was exposed from under its sedimentary veneer. This is an important idea to keep in mind, because whatever was being eroded from the highlands could show up as debris carried by the rivers that drained them. The most compelling lines of evidence for the existence and location of the Mogollon Highlands are the absence of strata today in that area to the south and the gravel and sand that were washed out of them.

The Granite Dells near Prescott are a remnant of the Mogollon Highlands southwest of Sedona

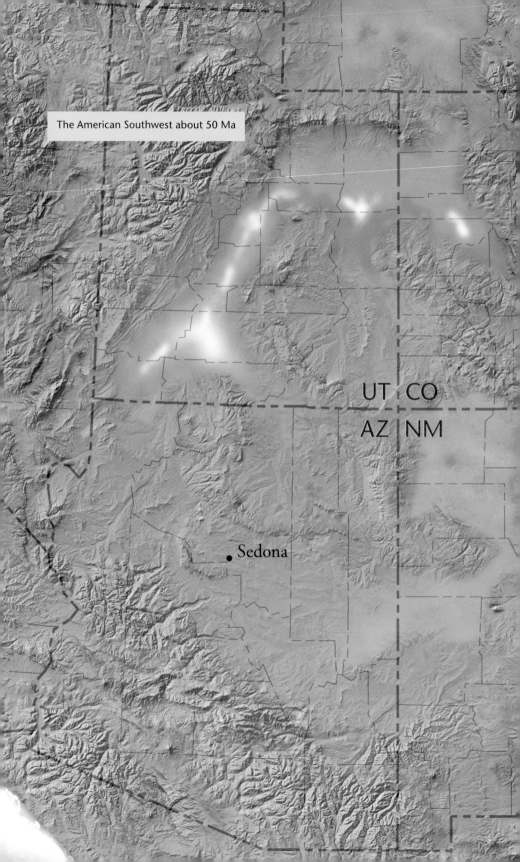

The American Southwest about 50 Ma

UT CO

AZ NM

• Sedona

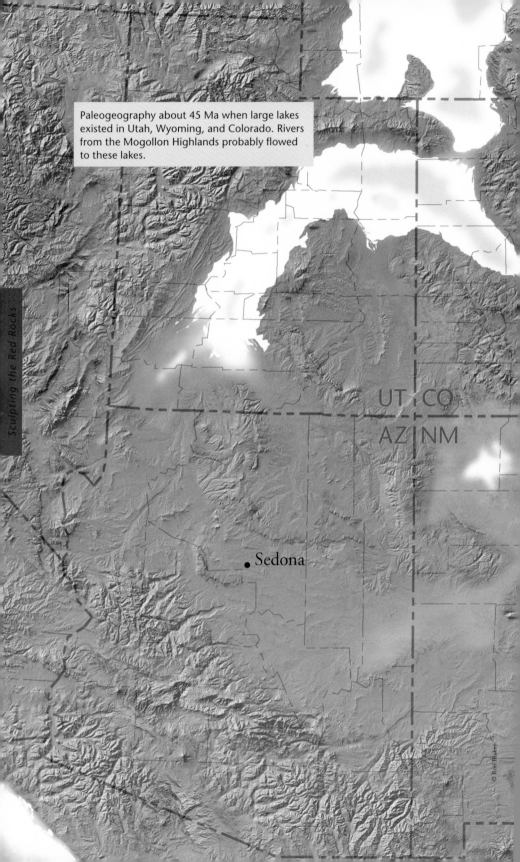

Paleogeography about 45 Ma when large lakes existed in Utah, Wyoming, and Colorado. Rivers from the Mogollon Highlands probably flowed to these lakes.

UT CO
AZ NM

. Sedona

© Ron Blakey

ANCIENT RIVERS:
The Formation of the Mogollon Rim
50 to 20 Million Years Ago

The uplift of the Mogollon Highlands caused ancient rivers to flow northeast towards Sedona and the Colorado Plateau, both of which were still lower in elevation than the highlands. Some of these rivers left gravel deposits and although very little of it remains on today's landscape, what is found is inordinately important because it reveals how the landscape evolved through time. Ironically, although these gravels are quite restricted in their exposure, they generate some of the most hotly debated ideas regarding Sedona's geologic story. Some geologists are baffled by their difficult-to-detect age and how they may relate to the other landforms that were emerging at this time. Different theories have been proposed about how and when the gravels were deposited, and an examination of all of these possibilities is instructive. An open-minded examination of the gravel story may help to resolve their puzzling origin.

Geologists agree that the oldest preserved deposits from rivers that flowed out of the Mogollon Highlands belong to an informal unit called the Rim gravel (by convention, rock units that are not yet formally described are placed in lower caps, i.e. Rim gravel is written in this way). It is composed of well-rounded and well-sorted clasts of **Paleozoic** sedimentary and Precambrian crystalline rock types. Since rocks like these were only being exposed to erosion south of Sedona, that is the direction

An exposure of the Rim gravel along State Highway 87 about 35 miles southeast of Sedona on top of the Mogollon Rim

from which the gravel must have come. The well-rounded texture of the clasts and the associated bedding features indicate deposition by rivers or streams that were flowing towards the northeast. The Rim gravel then, together with the nearly complete lack of Paleozoic and Mesozoic sedimentary rocks in central Arizona provides the key evidence that the Mogollon Highlands once stood in a lofty position southwest of Sedona.

Fossils have not been found in the Rim gravel making it very difficult to date the deposit precisely. Near Show Low, Arizona, a volcanic cobble dated at about 54 Ma was found within the gravel, so the deposit must be younger than this. At another location, a lava flow dated at about 28 Ma covers the deposit, so it must be older than that. Other outcrops of northeast-directed river gravels are found on top of the rim to the east of Sedona and these are more precisely dated between 37 and 33 Ma. This is most likely the age of the Rim gravel as well.

About 50 to 30 Ma, great interior lakes were present in what are now the states of Utah, Colorado, and Wyoming (see map on page 72). These lakes left behind the famous tar shale of that area and the colorful lake beds exposed at Utah's Bryce Canyon, now eroded into fanciful hoodoos. The Mogollon Highland streams that deposited the Rim gravel may have been a source for some of the water in these lakes. However, no outcrops of Rim gravel have been found to be interbedded with the lake deposits. Their inferred relationship is based solely on their similar ages and the known direction of stream flow from the highlands towards the lakes. Modern settings that have been proposed as analogous to the Rim gravel are found on the High Plains of eastern Colorado and western Kansas where large meandering rivers deliver loads of gravel and sand eroded from the Rocky Mountains.

The isolated outcrops of the Rim gravel are merely remnants of a much more widespread and thicker deposit. There is possibly only one exposure of it near Sedona on the Mogollon Rim and that is above Sycamore Canyon to the west. How much of the Rim gravel was eroded away from here we cannot know. But its cryptic presence reveals that rivers once flowed from a highland in central Arizona towards the northeast, and across a surface now capped by the Kaibab Formation on the Mogollon Rim. Its placement there strongly suggests that the rim did not exist during Rim gravel deposition. This pre–Mogollon Rim landscape existed before 30 Ma. The evidence documenting when the rim appeared on the landscape is obscure but another inconspicuous gravel may tell the tale.

Here the story becomes interesting. Below the Mogollon Rim and at the southern limit of Sedona's red-rock country, other deposits of river gravel are found. These are exposed along State Highway 179 south of Sedona, near a feature called Beavertail Butte. At first glance, these deposits look identical to the Rim gravel, being composed of well-rounded and well-sorted clasts of Precambrian crystalline rocks with a few well-rounded Paleozoic clasts thrown in. Their texture also suggests that they originated in river systems that only could have come from the Mogollon Highlands in the south. Some geologists have noted the similarities between these and the Rim gravel and prefer to use the same label for both. They explain the differences in their elevation in a couple of different ways (see the sidebar, "Origin of Ancient River Gravels" on pages 78 to 80).

Southwestern paleogeography about 30 Ma when the Mogollon Rim began to develop in central Arizona. Note the large dune field in the Four Corners area that existed at this time.

UT CO
AZ NM

Sedona

Yet, subtle but important distinctions have been noted between the two. Foremost among these is the substrate that underlies the two gravels. As noted previously, the Rim gravel is found only on top of the Mogollon Rim and is underlain always by the Kaibab Formation. The gravels near Beavertail Butte, however, rest on a much lower surface cut into the Hermit and Schnebly Hill formations and about 2,500 feet lower that the rim surface. This important distinction leads to the likely possibility that the Beavertail Butte formation, as it is informally called, is a separate though admittedly similar-looking gravel.

The Beavertail Butte formation contains three different members. The lower member is rich in angular, poorly sorted clasts derived from the Kaibab Formation and Coconino Sandstone. The Coconino clasts could not have traveled very far because they are not durable in transport. These gravels appear to be deposited

SE   NW

© Wayne Ranney

Close-up view of the Beavertail Butte formation. Note the orientation of the clasts suggesting deposition to the southeast.

against a slope of underlying **bedrock** that is cut into the Hermit and Schnebly Hill formations—a slope that I have interpreted as being an ancient part of the Mogollon Rim. The middle member is a fine-grained marl (claystone) that appears to have been deposited in shallow lakes beneath the ancestral rim. The upper member of the formation is a river deposit that could only have come from the Mogollon Highlands. This gravel, although initially transported to the north, shows evidence locally of flow towards the southeast, again suggesting that it was being deflected in that direction by the trace of the Mogollon Rim. This curious bit of evidence may also document the initial appearance of the Verde River in the area.

Exposure of the Beavertail Butte formation at its type section on Beavertail Butte along State Highway 179

Since the Rim gravel was deposited on top of the Mogollon Rim, and presumably before the escarpment came into existence, and the Beavertail Butte formation was deposited at the foot of the Mogollon Rim after it formed, the age of the two gravels can constrain the timing of the rim's formation. The Beavertail Butte formation contains a 28-million-year-old volcanic clast, meaning that it must be younger than this. It is also capped by a 15-million-year-old lava flow at House Mountain and so must be older than that. Thus, its age is between 28 and 15 Ma with an average age likely between about 25 and 20 Ma. Recall that the best age estimate for the Rim gravel is between 37 and 33 Ma. These constraints suggest that the Mogollon Rim likely appeared on the landscape sometime between 33 and 25 Ma and probably formed by the down-dip erosion of strata away from the Mogollon Highlands or, conversely, by incision into the strata south of Sedona by an ancestral Verde River.

The Beavertail Butte formation is easily overlooked being placed among the captivating red rocks, but these subtle outcrops most likely document when the Mogollon Rim or even the Verde River came into existence. Although the exposures are not as spectacular as those of the Schnebly Hill Formation, the story they preserve is no less impressive. Walking in these inconspicuous, rounded hills, one can almost hear the trickle of an ancient river flowing beneath a cliff of towering red sandstone.

ORIGIN OF ANCIENT RIVER GRAVELS

One of the outstanding questions remaining in the interpretation of Sedona's geologic history is how ancient river gravel came to rest both on top of, and at the base of the Mogollon Rim. And because the rim is over 2,500 feet high this discrepancy in elevation is no small matter. Practically none of the so-called Rim gravel remains above Sedona and not much more of it (the Beavertail Butte formation) is left below. There is a complete lack of fossils within these gravels and a paucity of other datable materials. Nonetheless, a proper interpretation of their origin would allow for a precise understanding of how and when the rim was formed.

Very isolated outcrops of river gravel have been found on top of the Mogollon Rim but mainly to the east of Sedona. It is assumed however, that the deposit once covered much of the rim surface in all of northern Arizona. There is a bit more gravel found below the rim in the Sedona area but this too is not widely preserved. Both gravels contain clasts that could have originated only from the southwest – even today, long after gravel deposition, southwest is the only direction where that type of rock is exposed to erosion. What is perplexing to geologists is that the presence of the river gravel on top of the rim implies gravel deposition before the rim came into existence (how else could it "climb" over it), while its presence below implies deposition after the rim had formed (how else could it come to rest there). Three intriguing alternatives have been formulated to explain the puzzling presence of the gravels above and below the Mogollon Rim.

The first idea maintains that much of the modern landscape elements in northern Arizona formed between 70 and 40 million years ago. According to this view, the Mogollon Rim (as well as the Grand Canyon, for that matter) was in existence at that time and looked much as it does today. This proposal suggests that deposition of the gravel began with initial accumulations on the valley floor beneath the rim, that gradually became thicker and thicker until the rim was completely buried and overtopped in gravel deposits. Later erosion of this thick pile is what caused the deposits to become separated.

Rim gravel

Rim gravel

Rim gravel remnants

Rim gravel remnants

© Bronze Black

1st Alternative – The ancient Mogollon Rim is buried in gravel. Later erosion separates the deposit.

A second alternative is that the gravel on top of the rim was deposited first as a broad sedimentary sheet across a beveled and rather featureless rim surface. When faulting formed the Verde graben between 8 and 10 Ma, the Rim gravel was faulted down along with the floor of the Verde Valley. Curiously, these first two proposals are in accord with one another in that they view the two gravels as a single deposit. Ironically, they disagree on which part of it was deposited first, the manner in how it became separated (erosion vs. faulting), and the role that the Mogollon Rim may have played in its origin.

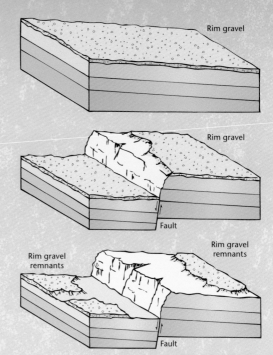

2nd Alternative – After widespread Rim gravel deposition, faulting and erosion separates the deposit.

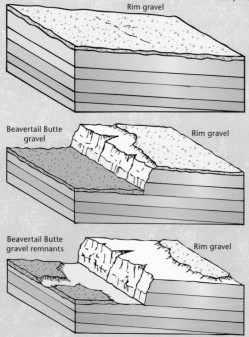

Preferred Alternative – Erosion creates the Mogollon Rim after Rim gravel deposition. The Beavertail Butte formation is deposited at the foot of the rim.

There is a third alternative, which I favor. I made a detailed geologic map of the gravels below the rim near Beavertail Butte and discovered that this gravel is different from the Rim gravel in subtle but significant ways. First, the valley gravel showed an ultimate transport direction to the southeast (not northeast like the Rim gravel). Additionally, significant faulting between the valley gravel and the top of the rim was not found. Any combination of faults would have to show at least 2,500 feet of offset to support alternative two, but only about 400 feet is detected. Lastly, the valley gravel clearly pinches out depositionally against a rising slope of Paleozoic bedrock to the northeast. In other words, there is no possibility that this gravel could have been deposited farther to the northeast because layers of bedrock in that direction prevented it from doing so. Consequently, I made a determination that the two deposits are different and proposed an informal name, the Beavertail Butte formation, for the valley gravel.

In 2001, Dr. Richard Holm of Northern Arizona University seemed to verify alternative three by completing a detailed study of the gravel clasts in both deposits. He showed that the Rim gravel contains proportionally more clasts of younger rock types, while the Beavertail Butte formation contains proportionally more clasts of older rock types. This provides evidence for the gradual unroofing of the Mogollon Highlands, whereby the youngest rocks in the source area were eroded first to become entrained in the Rim gravel, while the older rock clasts were eroded later and deposited in the Beavertail Butte formation. This is a subtle distinction not readily discerned in casual observation. Holm also noted that the Rim gravel is capped to the east of Sedona by a 28-million-year-old lava flow (meaning that it is older than this), while the Beavertail Butte formation contains a 28-million-year-old volcanic clast (meaning it is younger than this date). Taken together this evidence seems to disprove that the two deposits are one in the same.

As geologists reflect on the pros and cons of the three alternatives, the crux of the dilemma boils down to this: how much paleotopography was in existence on the landscape near Sedona before faulting related to the Verde graben commenced about 10 Ma? All three proposals spiral in various ways towards this essential question. Alternative one says that all of today's topography existed well before 10 Ma, although much of it may have been buried by gravel for a significant interval of time. Alternative two says that none or at most, very little topography existed before 10 Ma and that all that we see today was formed by faulting and subsequent erosion. These two alternatives are mutually exclusive and cannot both be true.

I favor alternative three because of the evidence described above but also because it has the distinct advantage of incorporating the most important aspects of the other two. Numerous independent studies, including my own, show that ancestors to the Verde Valley and the Mogollon Rim first came into existence between about 33 and 25 Ma. Although the ancestral Mogollon Rim may have had a more subtle profile than the sheer cliff we see today, portions of it existed nevertheless (see the sidebar on page 88 for a discussion). The preferred alternative shows how deposition of the gravel in the Beavertail Butte formation was influenced profoundly by an early topography, but was later modified or "overprinted" by faulting when the Verde graben became active. Thus, the preferred alternative is inclusive of the most important aspects of the other two: that the Mogollon Rim was in place before the latest gravels were deposited and that faulting and erosion put the finishing touches on all of it. While science is never practiced with regard to notions of democracy, good science neither is ignorant of the facts for all the types of evidence.

Origin of Sedona's Gravels

N   S

Flagstaff Sedona Jerome Prescott Phoenix

100 Ma

Mesozoic Rocks (Mz)

Paleozoic Rocks (Pz)

Precambrian Crystalline Rocks (Pϵ)

A pre-uplift, pre-Mogollon Rim cross section of central
Arizona from Flagstaff to Phoenix

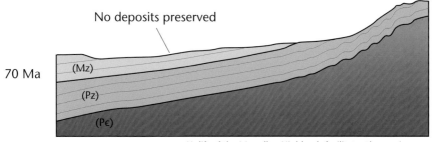

No deposits preserved

70 Ma

(Mz)

(Pz)

(Pϵ)

Uplift of the Mogollon Highlands facilitates the erosion
of strata in that part of central Arizona

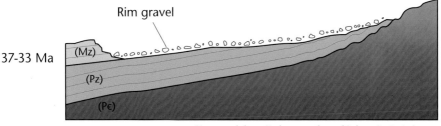

Rim gravel

37-33 Ma

(Mz)

(Pz)

(Pϵ)

Rim gravel deposited on a surface of Paleozoic strata in
central Arizona

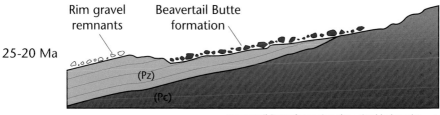

Rim gravel remnants Beavertail Butte formation

25-20 Ma

(Pz)

(Pϵ)

Beavertail Butte formation deposited below the
Mogollon Rim in the Sedona area

Paleogeography about 15 Ma when volcanic activity associated with the Hickey Formation was prevalent in central Arizona. This is when the House Mountain volcano was erupting.

UT CO

AZ NM

Sedona

©Ron Blakey

House Mountain:
SEDONA'S VOLCANO
15 to 13 Million Years Ago

Virtually unknown to most visitors to Sedona is a beautifully preserved volcano that exists within seven miles of the city center. The volcano's gentle slopes, dotted with evergreen pinyon and juniper trees, shrink to scenic insignificance and are lost among the grandeur of the red rocks. But visually appealing rocks are only part of the great geological heritage on display in this high-desert wonderland. It is interesting to note that the official Scenic Highway designation along State Route 179 south of the Village of Oak Creek ends abruptly when the last outcrop of red sandstone disappears. But the area's scenic geology doesn't end south of Sedona—it just doesn't sell as well. Geologists know that scenery is a continuum and to them a boring drive, especially in Arizona, is a contradiction in terms. To put it simply, if the red rocks are the cake in Sedona's geology, then the House Mountain volcano is the icing.

When told that a volcano exists south of Sedona, most people probably expect to see something like Mount Fuji or Mount St. Helens standing tall in the Verde Valley. But House Mountain is a **shield volcano**, a form that typically has gentle, low-lying slopes. Shield volcanoes form when very fluid lava erupts and flows for many miles in all directions away from a central vent. When cooled, this lava forms a rock called basalt, which, incidentally, is the most common rock type found on our planet. The

© Wayne Ranney

The House Mountain shield volcano as seen from near Cornville

basalt slopes of shield volcanoes are generally inclined at less than 15 degrees and this quality alone makes them difficult for an untrained eye to discern. Perhaps this is why the secrets of House Mountain were only discovered in the late 1980s. Erupting within sight of colorful red rocks is no way to achieve notoriety. For a long time House Mountain was the domain of only the eagle, the mountain lion, and the rancher.

When geologists first began to study the volcano, it was originally thought to be only about 5.5 million years old. This was based on a date that was obtained from a

lava flow exposed on its south side. When detailed mapping of House Mountain began it became apparent that this date might be too young. Gullies on the slopes of the volcano contained deposits that were between 6 and 8 million years old, making it impossible for the volcano to be just 5.5 million years old. Further mapping revealed that the 5.5–million-year-old lava flow had come from some source other than House Mountain. Two samples were then collected from near the top of the volcano and these gave ages of 14.5 and 13.2 million years. The volcano

Aerial view towards the southwest of House Mountain

was nearly three times older than previously thought.

More confusing relationships were discovered as mapping continued. The volcano's broad, gentle slopes extend a fair distance from the central vent but only in three directions—east, south, and west. Curiously, no lava flows were observed to the north where layers of red rocks are exposed in a valley near Turkey Tank and the Village of Oak Creek. Recall that some scientists believe that the Beavertail Butte gravel is a down-faulted portion of the Rim gravel and it was first assumed that the "missing" lava flows on the north side of House Mountain might also have been faulted and eroded away. That answer could only be found by locating a place where the basalt is in direct contact with the red rocks (see photos on pages 85, 88, and 137).

These outcrops however, are difficult to find because weathering of the basalt creates a loose soil that obscures the contact. Scrambling and bush-whacking through some of Sedona's least-traveled areas eventually uncovered a small but very important outcrop that provided a window through which the story of the volcano's origin could be seen. At this outcrop, just east of the summit trail from Turkey Tank, a layer of volcanic scoria (cinders) was found in contact with the red rocks. The presence of scoria is informative because oftentimes volcanic eruptions begin with gas-infused lava that is hurled into the air as fiery fountains. As the droplets of hot lava cool and fall back to the earth, they form scoria deposits. This outcrop most likely preserves the material initially erupted from House Mountain.

Not surprisingly then, there was no evidence for faulting between the scoria and the red rocks. Instead, the scoria was found lying on an eroded surface carved into the red rocks. The contact did not show the ivory-smooth surfaces called **slickensides** that form when movement along faults grinds rock surfaces together under pressure. Rather, the texture of the red rocks here was as rough as sandpaper. Additionally, the faults in the area dip to the southwest at an angle of about 75 degrees or more, but the contact between the scoria and the red rocks sloped only about 55 degrees. This relatively gentle dip and the lack of slickensides indicated that House Mountain was not down-faulted against the red rocks, but rather was in depositional contact with them.

But questions remained. Why were there no lava flows to the north of House Mountain? How did the area known as Big Park (where the Village of Oak Creek is built) escape being filled with scoria and lava, especially since it was located so close to the central vent?

The relationships exposed at this outcrop finally unraveled the volcano's origin. The lack of lava north of House Mountain was the result of volcanic eruptions that

View to the east from the Turkey Creek trail showing the contact between House Mountain lava and sedimentary strata. This contact (dotted line) shows the slope of the Mogollon Rim about 15 to 13 Ma.

occurred below the Mogollon Rim. The rim formed a barrier that prevented lava from flowing northward. What we see today is an example of **inverted topography**, whereby something that was emplaced low on the landscape (House Mountain volcano) is now higher in elevation because of erosion of the sandstone around it. About 15 to 13 Ma, the Mogollon Rim was the high feature on the landscape and the House Mountain volcano erupted in a valley below it. The volcano's durable basalt has been much more resistant to erosion than the sandstones that made up the slope of the rim. This inverted topography can be seen on the skyline southwest of the Village of Oak Creek, where a long, basalt-topped mesa is exposed. Its outline traces the location of the Mogollon Rim about 13 Ma (see photo on page 121).

I once believed that the relationship between the House Mountain lava and the ancient location of the Mogollon Rim could allow for a calculation of the average rate of retreat or erosion of the rim through the last 13 million years. I did this by measur-

© Wayne Ranney

Close-up view of the boundary between the ancestral Mogollon Rim (left) and scoria from the House Mountain volcano. The pencil lies against the old rim surface.

ing the distance from the north edge of the lava flow (where the rim used to stand), to the nearest cliffs of the modern rim on Lee Mountain. This distance of about four miles, when divided by the age of the youngest lava, gave an average rate of retreat of about one foot every 625 years. I thought that this rate was further verified by evidence from another lava flow capping Horse Mesa south of the Village of Oak Creek. This flow is dated at 6.4 Ma and is now about two miles from Lee Mountain. A quick calculation from this location gives an average rate of retreat of about one foot every

606 years. These nearly identical rates of retreat from two separate sources seemed to verify the notion.

However, in the twenty years or so since I made this initial interpretation, I have developed new ideas about how the Mogollon Rim has evolved through time. I still believe that House Mountain volcano erupted below the rim, but I and at least one other local geologist believe that the rim may have had a more gentle slope previously, giving it a much more subdued profile than we now see. This ancient rim probably looked different than the near-vertical wall present today and if so, the manner in which the rim has evolved since House Mountain time necessitates different processes in its formation than that of a steep cliff retreating steadily to the northeast.

In any event, the eruption of the House Mountain volcano must have been a spectacular sight. Imagine standing on top of the rim and watching fiery fountains of lava shoot high into the air, and flows eventually building a broad shield volcano below.

The Mogollon Highlands were most likely gone by this time because about 17 Ma the earth's crust underlying it had stretched and thinned, creating the Basin and Range topography exactly where the mighty highlands once stood. On a distant continent named Africa, a species of primates was still living in the trees and eating leaves and nuts, still 10 million years away from an experimental life on the grasslands. Yet House Mountain would live and die, become totally buried in lake sediments, and then be completely exhumed before the first pair of human eyes saw its slopes only about 13,000 years ago.

To a species interested enough to ask, the House Mountain volcano finally yielded its secrets. However, this long-lost landscape did not suddenly appear in a single flash of insight. It emerged gradually, over a period of many months, and only by constant inquiry and the testing of all possibilities. Each successive journey into the field revealed a clearer picture of how the volcano came to be. When a person finally visualizes a landscape that has never before been seen, the sense of discovery that fills the soul is a feeling that cannot be described adequately in words alone. It transcends scientific inquiry to become a new way of perceiving time and our place within it. To know a piece of land, to understand a landscape, this is what it means to belong to the living earth.

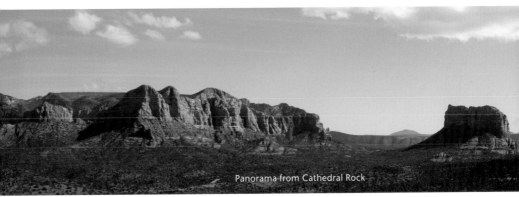

Panorama from Cathedral Rock

© Bronze Black

EVOLUTION OF THE MOGOLLON RIM

While conducting research and geologic mapping in the House Mountain area of Sedona in the 1980s, I helped to develop an interpretation of how the Mogollon Rim had evolved. This idea—that the rim was created by the gradual down-dip retreat of strata to the northeast, away from the uplift of the Mogollon Highlands—is still well founded. However, in the last few years it has become apparent to me and to fellow geologist Ron Blakey, that the rim might have had a somewhat different appearance than what I originally envisioned more than twenty years ago. Although this new concept and the research it involves is still developing and is not yet formally accepted, it's a thought worth sharing to explain how new geologic ideas emerge and are gradually developed into theories.

Today the Mogollon Rim is a massive, nearly sheer-walled escarpment that towers more than 2,500 feet above the city of Sedona. When I found that the House Mountain volcano had erupted below an ancestral alignment of the rim over 13 Ma, I naturally assumed that this escarpment had a profile similar to the one observed today. But if geologists are lucky enough to "hang around" and look critically at a particular landscape through decades of time, they are able to pick out subtle nuances within that landscape. Such observations are never possible in the first years of one's experience with a landscape, no matter how much experience they have elsewhere. Only time and repeated exposure to something allows for this type of intimate view. Both Ron Blakey and I have independently begun to see the same thing in Sedona's landscape, and the independent nature of our observations is a compelling pull for both of us to continue to develop this idea.

We have no-ticed a subdued, gently eroded surface that is

perched between the top of the Mogollon Rim and many of the well-known buttes that stand below it. Examples of this older surface are preserved and can be seen on the tops of Courthouse Butte, the Seven Sisters, Cathedral Rock, Twin Butte, and Doe Mountain. The west side of Munds Mountain also displays this surface just below the rim. This gentle surface descends from the top of the Mogollon Rim and gradually runs down to the level of the Coconino Sandstone. It is a rolling topography, open enough to allow trees to grow on it, but is perched above the sheer exposure of Schnebly Hill Formation below it, where the floor of the modern landscape exists.

Remnants of this old surface extend downward through strata to the southwest and beneath the House Mountain volcano. This means that the ancient surface must be older than 15 Ma, which is the oldest age of those lavas. It is likely that the Beavertail Butte formation also rests on this surface and if so, then the surface is older than about 25 Ma. Preliminary investigations of the ancient surface suggest that the Mogollon Rim in Sedona was well established when the House Mountain volcano erupted and may have been in some early incarnation when the Beavertail Butte formation was deposited. During this time the rim had a much gentler profile than it does today. There were likely two steps in the ancestral rim, with an upper slope confined between the Kaibab and Coconino formations and a lower slope cut down through the Schnebly Hill Formation. Remnants of an extensive bench connecting these two slopes remain in places along the contact between the Coconino and Schnebly Hill formations. Only within the last 6 million years has a steeper profile of the modern rim been etched into the landscape. Look for remnants of this old surface on top of some of the buttes and spires as you explore Sedona.

© Wayne Ranney

A view of Twin Butte from the south. Its upper surface may be a part of an older terrain that existed in the early history of the Mogollon Rim.

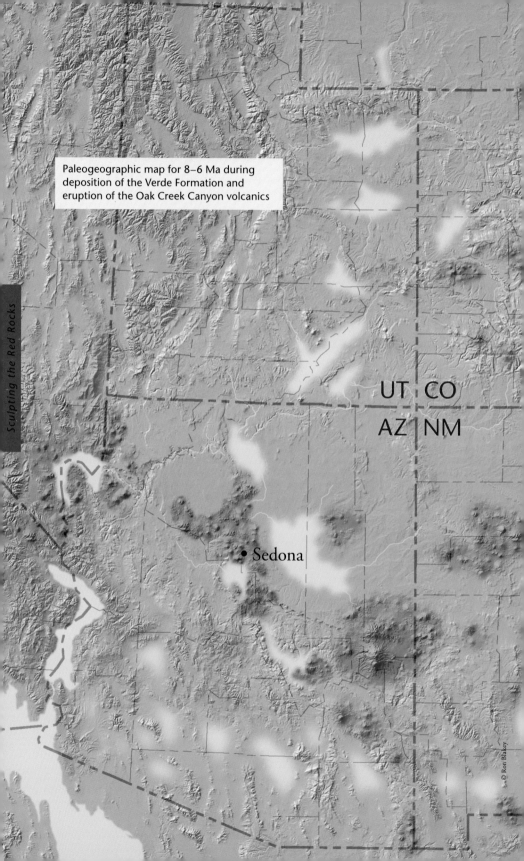

Paleogeographic map for 8–6 Ma during
deposition of the Verde Formation and
eruption of the Oak Creek Canyon volcanics

• Sedona

UT | CO

AZ | NM

Sculpting the Red Rocks

© Ron Blakey

The Verde Valley and Oak Creek Canyon:
CREATING THE MODERN LANDSCAPE
10 Million Years Ago to the Present

The last chapter of our story centers on the finishing touches that sculpted this landscape masterpiece. It involves forces that created the Verde Valley, a freshwater lake that filled the valley, and movement on the Oak Creek fault, which facilitated the carving of Oak Creek Canyon. All of the major landforms that bring people to this area by the millions originated within this relatively short period of time. And as we shall see at the end of the chapter, the geologic parade that has been moving through this area for hundreds of millions of years will not stop simply because we humans have now arrived. What we experience and enjoy in this marvelous terrain is but one frame in an ever-evolving movie. Even if we are not around to view the future frames, the show will go on.

Aerial view of Oak Creek Canyon and Oak Creek fault showing the offset between Wilson Mountain (upper left) and Wilson Bench (center)

The demise of the Mogollon Highlands was not simply due to their gradual erosion. Beginning about 17 Ma the earth's crust in the American Southwest began to stretch and thin. Whereas **compression** had previously uplifted the area, now it experienced a new kind of stress called **extension**. Broad areas of the former highlands were down-faulted creating the Basin and Range Province, and it was during this time that the region's streams began their slow reversal of flow from northeastward to southwestward. One of the faults related to this extension is the Verde fault, which uplifted the Black Hills and down-dropped the Verde Valley. Movement on the fault began between 10 and 8 Ma and coincidentally ripped through the crust right where ancient volcanoes had previously left deposits of copper and gold. In fact, the old mining town of Jerome was constructed directly on top of the Verde fault.

As the fault continued to slip, a tectonic basin developed and the modern Verde Valley began to take shape. Soon runoff from the Black Hills and the Mogollon Rim began to accumulate in the rapidly subsiding basin and an **ephemeral** lake was born. During wet periods the lake was quite extensive within the confines of the valley, but in drier times the lake would shrink or even dry up altogether. Near the town of Camp Verde, salt and gypsum deposits are found that document these evaporating conditions. In the central part of the valley over 3,000 feet of the Verde Formation accumulated during a 7- or 8-million-year period. Through time, the House Mountain volcano was buried in these deposits of white limestone (high lake water), red mudstone (low lake water), and salt and gypsum (evaporating lake water). Some very large animals lived along the shores of this lake including stegomastodons, camels, and tapirs. Nearby attractions such as Montezuma Castle, Tuzigoot, and Montezuma Well are found within the white limestone of the Verde Formation.

Obviously, Sedona is not known for white limestone but rather for its dazzling red-rock formations. And the verdant beauty found within the confines of Oak Creek Canyon adds a sylvan, almost elfin quality to the landscape. Nearly as world-renown as the Grand Canyon to the north, Oak Creek Canyon is a watery wonderland of graceful sycamores and lush ferns all set beneath the stunning red rocks. The carving of this canyon deep into the face of the Mogollon Rim involves many of the same processes that created other nearby canyons such as the Grand Canyon. But for a long time the story of how and when Oak Creek Canyon formed was difficult to sort out because of the confusing relationships between the lava flows found within it and the Oak Creek fault. Fortunately, detailed studies of the basalts and the fault have made the story clearer. The specific placement of the canyon is not the result of random erosion back into the rim, but was determined by the location of the fault and forces that predate the cutting of the canyon by hundreds of millions of years.

The canyon's more recent history began when the same pressures that uplifted the Mogollon Highlands affected rocks here. These forces of compression caused the earth's crust to buckle and break along fault lines that perhaps originated at a much earlier time. Once faults form, they can remain as active lines of weakness throughout the rest of geologic time. Recall that the oldest crust in the area is found in the Black Hills near Jerome, and when those rocks were initially squeezed and fractured a regional system of faults was established. The Oak Creek fault may represent reactivation of an older fault line that was for some time covered by many thousands of feet of sedimentary rocks. The fault is about thirty miles long and generally trends north-south from an area just south of the San Francisco Peaks to Horse Mesa, east of the Village of Oak Creek.

About 80 Ma the Oak Creek fault experienced movement again. Compression caused the rock layers to bend at first, then fracture and break when the pressure became too great. The bending of the red rocks can be seen along Highway 89A, about one-third of a mile north of Grasshopper Point. The inclined strata reveals that this episode of movement shoved the rocks east of the fault almost 600 feet higher than those on the west side. On the eastern block, uplifted strata belonging to the Kaibab, Toroweap, and Coconino formations were eroded away. This explains why the east side of Oak Creek Canyon does not have these rock layers today. The uplifted eastern block eventually eroded down to about the same level as the top of the western block. This surface was planed and presumably buried by the Rim gravel. After these events, a broad valley was cut to the east of the fault and into the edge of the emerging Mogollon Rim.

Between 8 and 6 Ma, basalt lava flows spilled from the top of the rim and partially filled the wide valley. These basalts can be seen in the eastern wall of Oak Creek Canyon, from the top of the switchbacks at Oak Creek Vista all the way to Horse Mesa south of the Village of Oak Creek. Their top surface is the gently inclined ramp that Interstate 17 follows between Flagstaff and Camp Verde. The

Basalt flows capping red rocks in Oak Creek Canyon

lava flows originated from **fissures** near the Mogollon Rim and flowed southward. A narrow basaltic dike that fills one of these fissures is located across the highway from the Junipine Resort in Oak Creek Canyon. And a cinder cone on top of Wilson Mountain is part of this same volcanic event. The wide valley filled with about 500 feet of basalt, resulting in a very subdued landscape.

But the topography of the area began to change drastically beginning about 6 Ma, when the Oak Creek fault became active again. Unlike the previous episode of faulting where the east side was raised, this time the east side moved down. The cinder cone on Wilson Mountain was cut by the fault and dropped about 830 feet to form Wilson Bench. A parallel branch of the fault became active east of Wilson Bench with a displacement of about 160 feet, making for a total displacement on the fault of about 1,000 feet. Looking to the north from the Midgley Bridge parking area one can see the main branch of the fault clearly as it offsets the lava flows on Wilson Mountain and Wilson Bench. Oak Creek cut its channel along the other branch of the fault between Sliderock and a place near Grasshopper Point. The two branches merge as the fault turns to the east for one mile, before trending again to the south along the far side of Munds Mountain where the fault terminates beneath Horse Mesa. Interestingly, Munds Mountain belongs to the same uplifted block of rock as Wilson Mountain. What separates the two is simply the later incision of Oak Creek into the landscape.

Along much of its length Oak Creek flows parallel to the fault. It is not uncommon for streams and canyons to follow faults—they are lines of weakness that water can attack and erode readily. So, why does modern Oak Creek depart from the fault near Grasshopper Point rather than following its former valley on the east side of Munds Mountain? One possible explanation requires us to use a perspective seen from atop Airport Mesa in Sedona, looking up into the mouth of the canyon. From this vantage we can look into the canyon and see the down-faulted basalt across the Oak Creek fault. When deciphering this enigma, it helps to remember that streams evolve, often expanding in the upstream direction by a process known as **headward erosion**. How could headward erosion explain why Oak Creek does not follow the entire length of the Oak Creek fault?

Remember that an ancestral valley existed once to east of the Oak Creek fault and Munds Mountain. When this valley filled with basalt lava flows, melted snow from the plateau infiltrated the ground and was funneled in the subsurface along the fault. At the same time, another drainage may have existed to the south in the vicinity of uptown Sedona with a southwest-facing **headwall** in red sandstone. Groundwater flowing along the fault line most likely percolated away from the fault as it turned east near Grasshopper Point, infiltrating the red sandstone and producing springs in the headwall of southern drainage. By a process known as **groundwater sapping** (where springs weaken and erode the rocks out of which they flow), this headwall was under-cut and collapsed progressively such that headwall erosion in this drainage proceeded north back into the fault line.

Ultimately, this drainage lengthened its channel in the sandstone and intersected the Oak Creek fault near Grasshopper Point. Once this happened, the drainage

The Oak Creek fault has offset Wilson Mountain (on the skyline) from Wilson Bench (center)

captured the subsurface groundwater flow along the fault and caused Oak Creek to lengthen its channel to the north along the fault line. This explains how upper Oak Creek Canyon was formed and why it does not follow the fault south of Grasshopper Point.

The deep dissection of the canyon is a relatively recent phenomenon. Evidence from the Grand Canyon suggests that cutting of the deep canyons on the Colorado Plateau did not begin until after 6 Ma and that much of it has occurred in only the last 2 to 3 million years. An interesting deposit from Oak Creek about this age is found on top of Airport Mesa, also know as Table Top Mountain. Some intrepid pilot realized that this apparently flat mesa would be a good place to land airplanes, where he wouldn't have to clear many rocks or other obstructions. The pilot probably didn't know that Oak Creek was responsible for creating this great landing strip.

Table Top Mountain is capped with a 100 foot-thick deposit of cobbles, pebbles, and dirt. The cobbles and pebbles are composed of rounded basalt, with only an occasional clast of red sedimentary rock. This composition is identical to that found in Oak Creek today—the same size, same **sorting**, and same composition of clasts. The similarities do not end there. Although the mesa appears flat, it actually slopes to the southwest at a gradient of about eighty feet per mile. The gradient in modern Oak Creek is about seventy feet per mile. The deposit is narrowest to the northeast, where it is closest to the mouth of Oak Creek Canyon, and widens gradually to the southwest. The deposit averages one and one-half miles long and a half mile wide. All of this information leads to the conclusion that Oak Creek probably deposited this sediment, even though the creek now flows 700 feet below Airport Mesa.

HOW WAS OAK CREEK CANYON CARVED?

The American Southwest is a land of stupendous canyons, and many studies about their formation have originated here. How canyons are carved and deepened is a fascinating topic that has occupied the attention of a number of local geologists. As a student, I first learned that canyons were formed by the slow, inexorable wearing away of the bedrock channel by muddy water. The idea was that sand and mud, carried by moving water, scoured like sandpaper. But Oak Creek is usually not a muddy stream. If muddy water is not responsible for cutting of Oak Creek Canyon, what is?

As it turns out, it is the relatively rare but large floods that race down its channel after a cloudburst that chisel down into the bedrock channel. And surprisingly perhaps, it is not the water in floods that does the excavating but rather the material that the water carries with it—namely, the cobbles and boulders that are washed into the stream from the cliffs above. These roll forcefully with the floodwater, scouring out sections of the bedrock channel. Many of these boulders in Oak Creek are composed of the hard basalt from lava flows on the rim above, and these durable clasts are excellent excavators.

It is not just floods in the modern era that have contributed to Oak Creek's depth, but also those

View of Oak Creek Canyon north towards the San Francisco Peaks

in ancient times when the climate was wetter in northern Arizona and the floods were larger. Such was the Pleistocene Ice Age, which ended only about 10,000 years ago. There was never any glacial ice in Oak Creek Canyon, but periods of wetter climate may have contributed to increased runoff that filled the channel of the stream more often.

If we knew the exact age of this deposit, we might be able to determine the rate at which Oak Creek is cutting its canyon. Unfortunately, no fossils have been found within this sediment to help determine its age. (Horse teeth or elephant remains would be the most likely fossils encountered.) However, logic and a few clues can be

© Wayne Ranney

Basalt boulders found on top of Airport Mesa are identical to those found in the modern bed of Oak Creek. This strongly suggests that Oak Creek deposited these boulders here approximately 2 Ma.

used to approximate its age. The gravel must be younger than the Verde Formation since there are no white limestone deposits in it. Its minimum age is about 2.5 Ma and a likely date for the deposit on Table Top Mountain is about 2 million years. When we divide that amount of time by the deposit's elevation above the creek, we get an average rate of cutting equal to one foot every 2,850 years, or about 350 feet every million years. This corresponds with studies done on the Little Colorado River on the Colorado Plateau, which cuts 320 feet every million years.

All of the observations above are stunning in what they reveal and show that Oak Creek Canyon has experienced a varied and interesting history. It shows evidence of faulting that moved in at least two different directions at two different times, was eroded into a wide valley that was covered in gravel first and subsequently buried by lava flows. In just 6 million years the creek has carved the colorful defile we see today. It makes one wonder, what might the area look like at some time in the future?

The canyon will undoubtedly become deeper as Oak Creek carves into the strata that underlie the red rocks. Even though we cannot see these rocks today, they must be quite similar to those of the same age exposed near Jerome or in the Grand Canyon. There are some differences in rock type and thickness between these areas, so only

HOW GEOLOGISTS DATE ROCKS

When geologists throw around large numbers for the ages of rocks and use terms involving millions of years, a natural reaction for many people is, "How do they know that?" It's a good question. The large numbers that geologists seem to be comfortable with are not wild guesses, nor are these numbers pulled out of thin air. Sophisticated techniques have been developed that allow scientists to know, within an acceptable margin of error, when a lava flow was red-hot or when an escarpment might have appeared on the landscape. And although it is not possible to directly date a sedimentary rock, a combination of dating and correlation techniques allow geologists to indirectly date rocks.

The most common dating technique is called radiometric dating, which involves measuring the ratio between different isotopes of specific elements that are trapped in rocks. Volcanic rocks are excellent specimens to date, although other igneous and metamorphic rocks can work just as well. When rocks like these cool, a clock starts ticking within them in the form of radioactive isotopes that naturally decay at a known and constant rate, transforming them into other isotopes. This rate of decay from one product (the parent) into another (the daughter) is a law of physics and cannot be altered or debated, although sometimes the amount of the original or by-product isotopes can be contaminated by other natural processes. In these instances reliable dates are not possible on this sample and another specimen must be obtained.

So let's say that a certain isotope common in volcanic rocks decays at the rate of 50 percent (it's half-life) every million years. That means that after one million years there will be 50 percent of the parent isotope and 50 percent of a daughter isotope. The parent will continue to decay at the same rate throughout subsequent time so that after another million years 50 percent of what was left at one million years will have further decayed. After two million years we will have 25 percent of the parent material and 75 percent of the daughter product. It's an ingenious way to date rocks—capture and measure the radioactive isotopes held within rocks and you can know when they formed.

The dates are usually given with a margin of error included. For instance, a lava flow in Oak Creek Canyon was dated at 5.99 + 0.14 Ma—almost 6 million years old or within 140,000 years on either side of this date. This means that we can say, with a 95-percent confidence level, that the flow erupted between 6.13 and 5.85 million years ago. Sedimentary rocks cannot be dated in this way but sometimes they may be interbedded with lavas or ash deposits that can be dated readily. Or they may contain fossils that have been found in other layers worldwide that might contain the datable ash. Through hundreds of years of correlation and many decades of radiometric dating, a network of age dates for sedimentary sequences has been determined.

Some non-scientists are skeptical of these radiometric dating techniques but it is not the role (nor the   intention) of the scientist to attack, discredit, or otherwise invalidate anyone's religious or spiritual beliefs. Geology helps us understand how the Earth came to look the way it does. Anyone who suggests that science is inherently antagonistic to religious thought is someone looking perhaps only for a controversy.

time will tell what the exact profile and color of a future Oak Creek Canyon will be. Based on the rate of downcutting documented from the Airport Mesa deposit, Oak Creek could expose the entire Paleozoic section and the very top of the Precambrian crystalline rocks in just a little over 4 million years (assuming 1,500 feet of sediment being cut at the rate of one foot every 2,850 years). One wonders if future landscape enthusiasts will be as lucky then to have a world-class copper deposit exposed in these rocks in the Sedona area.

The earth has been experiencing an ice age for about the last 2 million years; remember that everything called "civilization" has occurred only during a tiny 10,000-year warm interval. If patterns of the last 2 million years persist we can be relatively certain that colder, wetter conditions will return to the area. If this occurs, ponderosa pines will grow once again on the terraces where Sedona is built and Douglas fir and spruce trees will once again occupy Oak Creek Canyon. Of course, our burning of fossil fuels may, in the short run, have an opposite effect. Stay tuned for details.

We cannot know if human beings will exist to verify our generation's exploratory attempts to define the rate of change for a particular landscape. Just to know that the red rocks will endure, even if we are gone, is satisfaction enough. If nothing else, our study of this landscape has taught us that things do—and should—change. The red rocks existed long before we laid eyes on them and they will exist when the last human leaves this place. Time is infinite, and we are lucky to have experienced a glorious part of it. We are luckier still to have been able to see Sedona through time!

© Bronze Black

View from Capitol Butte

Flowering agave in the red rock country

THE HYDROLOGY OF OAK CREEK

It might seem odd to think of a lush, watery wonderland like Oak Creek in the high desert country of northern Arizona. But there are myriad groundwater resources here that make Sedona and the Verde Valley such hospitable places. The area's streams owe their existence to a fortuitous combination of geologic and climatic factors that have acted in concert through hundreds of millions of years to cause fresh water to emerge cold and clean from the ground. Ironically, these water resources begin with the deposition of arid-land sandstones such as the Supai, Schnebly Hill, and Coconino formations, whose porous nature contains enough space between grains to hold groundwater.

© Bronze Black

Snowfall is the source of spring water in the Verde Valley

The water originates on the Coconino Plateau, north of the Mogollon Rim, where rain and snow deliver between eighteen and twenty-two inches of precipitation annually. While most of this water evaporates back into the atmosphere or runs off as meltwater, it is estimated that about two inches of the total annual precipitation sinks into the rocks. This water finds its way into small cracks and crevices that interconnect with one another, ultimately finding its way into the sandstone aquifer. Layers of clay block the downward movement of the groundwater and the Oak Creek fault blocks its lateral movement. This combination causes water to emerge through springs in Oak Creek Canyon. We can think of the earth as like a giant sponge, and springs as leaks in the regional aquifer where the water table rises to intersect the earth's surface.

At the base of the switchbacks in the north end of the canyon Sterling Springs emerges to form the headwaters of Oak Creek. This spring releases only about 320 to 390 gallons per minute (gpm) but this is enough to operate the fish hatchery there. Over forty springs add to the total amount of water in all of Oak Creek and Sterling Springs contributes only about one percent of it. In an area just upstream from Indian Gardens, about 5,800 gpm is found in Oak Creek. Below Indian Gardens, large springs more than double the amount of water, which increases to about 15,700 gpm. It is no wonder the Yavapai and Apache people loved this place and used it as a garden. Curiously, below this point as the creek flows through Sedona to Page Springs, there is a net loss of surface water. Whereas the previous sections of the stream were "gaining reaches," this section is a "losing reach," such that the amount of water in the creek above Page Springs is less than 9,000 gpm.

The Page Springs area is replete with springs bearing colorful names: Lolomai, Bubbling Pond, Tree Root, Turtle Pond, and, of course, Page Springs. Within this limited reach, Oak Creek increases its flow to a whopping 28,700 gpm—a 220 percent increase. Below Cornville, Oak Creek flows into the Verde River after traveling about forty miles and dropping more than 2,600 feet, although it is still 3,200 feet above sea level at this point. From here the water travels down the Verde River through wild and rarely seen Sonoran Desert canyons, merges with the Salt and Gila rivers, and eventually enters the Colorado River at Yuma, Arizona, where it turns south and spills into the Gulf of California. Of course, in these modern times the water is all "used up" by the time it gets to the reservoirs above Phoenix and will likely never reach the sea at all. But what a journey our spring-fed creek has made from the forested Mogollon Rim to the mighty Colorado in Mexico.

A healthy future for this stream depends upon politically neutral studies of the regional aquifer and unbiased interpretations that will allow us to make informed decisions about how much growth our aquifers can support. In this way, we can live in harmony with the many species that depend upon this vital resource. Meanwhile, Oak Creek continues to inspire and draw people to its lush banks. For me, it's the smell of the sycamores, the rhythm of the cicadas, and the taste of spring water that will constantly keep me coming back.

© Wayne Ranney

Beautiful Oak Creek

View towards Snoopy Rock

PART 4

Where to See the

Red Rocks

Geology Road Logs and Hikes

© Bronze Black

GEOLOGY ROAD LOG 1
A Grand Geology Loop Around Sedona

This geology road log takes you on a scenic and informative drive from the heart of Sedona, around the House Mountain volcano, and back again in one giant loop. It includes great views and fascinating stories about the red rocks, faults, volcano's, springs, an ancient lake, lava flows, and back into to the red rocks. With eight stops that contain geology narratives, the drive has a cumulative distance of about 42 miles and will take you three to four hours to complete. The eight individual sections can be run as short

Highway sign in Oak Creek Canyon

mini-tours. Any time of the day is good for photography but some sections show better in the morning or afternoon light. Remember to pull over often to let speedy, non-geologists pass. Use turnouts whenever possible and use your hazard lights when driving slowly to take in a view. Please pay attention to the road first and geology second. Reset your odometer to 0.0 and you're ready to go. I wish you happy touring!

0.0   Begin the tour at the roundabout at Highway 89A and State Highway 179 in uptown Sedona. This junction was formerly called the "Y" but when the entrance to the Hyatt Piñon Point opened, the intersection took on the appearance of an "X." Completion of the roundabout in 2008 finalizes its metamorphosis as an "O." Go west and immediately after the "Y, X, O," you'll go straight through another roundabout at Brewer Road and the Sedona post office.

0.1   The Hermit Formation is visible in the roadcut to the right.

0.3   Crossing the Soldier fault with the southwest side sloping down.

Roadcuts in the Hermit Formation showing point bar deposits

0.4   Roadcuts in the Hermit Formation next 0.4 miles. Look for undulating beds of siltstone and sandstone, interpreted as **point bar** (words appearing in bold are defined in the Glossary beginning on page 152) deposits on the inside curves of ancient

104

stream channels.

0.8   On the left is a contact between the Hermit Formation and the overlying Schnebly Hill Formation. Get in the left lane at this time and prepare for a left turn at the Sedona Airport turnoff.

1.0   Turn left here and proceed uphill to the top of the mesa for a fantastic view of Sedona, the Mogollon Rim, and a geology discussion.

1.4   The Schnebly Hill Formation is exposed in a small roadcut on the left. The "airport vortex" is located on the hill ahead.

1.6   Pass the "airport vortex" (often crowded). As you make this final climb up to Airport Mesa, watch on the left for outcrops of the Schnebly Hill Formation, which eventually gives way to an inconspicuous but important basalt-rich cobble deposit.

2.2   Arrive at Airport Mesa Overlook. The parking area is on the left. Walk across the road for a view and a discussion of the geology from here.

Stop #1–
Discussion about the Geology from Airport Mesa Overlook

Welcome to one of the best views in all of Arizona! As you approach the overlook you will see the broad expanse of West Sedona at your feet with the Mogollon Rim

The view from Airport Mesa. Valley floor is carved in the Hermit Formation.

rising above it in the distance. The view extends from Black Mountain and Sycamore Pass far to your left (west) and to Munds Mountain much closer to your right (east). What created this landscape and what can be learned from here?

First, note the expanse of West Sedona below you. Before this area was urbanized, cattlemen grazed cows here and knew the area as Grasshopper Flat. This open valley formed because the relatively soft Hermit Formation erodes quickly (geologically speaking), and undercuts and collapses the harder layers above it. These harder layers are visible from here and include (from bottom to top) the red Schnebly Hill

Coffee Pot Rock and the "green line" on the Mogollon Rim

105

Formation, the slightly gold-colored Coconino Sandstone, and the white Toroweap and Kaibab formations. Try to locate the "green line" in the Mogollon Rim as you look up Soldier Canyon to the right of Coffee Pot Rock. This stratigraphic marker highlights the contact between the Coconino and Toroweap formations. Thin, resistant caps of black basalt may be seen on top of Wilson Mountain to your right.

© Bronze Black

Lava flows atop Wilson Mountain

This is a great place to discuss the geology of the Mogollon Rim. Only the Grand Canyon is a more important landform in Arizona. It is not evident from here, but the strata in this area dip slightly to the northeast and geologists have determined that this dip was imprinted in the rocks during the **Laramide Orogeny**, some 70 to 40 Ma. The dip developed as the Mogollon Highlands were uplifted to the south and west of Sedona near present-day Phoenix, Prescott, and Las Vegas (see the paleogeographic map on page 68). It is likely that the colorful rocks you see here were also present at one time in those areas to the south but were eroded from there at the time that rocks here were tilted to the northeast. Essentially, you are seeing only a single "frame" in a very long geologic movie. If we could see the movie in full, we would see the edge of the Mogollon Rim retreating to the northeast through time. Evidence from House Mountain volcano to the south suggests that the average rate of retreat is about one foot every 625 years.

Now look to the east towards uptown Sedona and you will see the mouth of Oak Creek Canyon opening into this broad valley. Note the dark wall of basalt exposed on the far side of the canyon. None of the colorful sedimentary rocks in the Mogollon Rim are found in this eastern wall of the canyon and in their place is 500 feet of black basalt. Through many years of observation, debate, and refinement of ideas, geologists have determined that an ancient valley was once present at this location, a valley that predated modern Oak Creek Canyon by at least 8.0 Ma. The lava flows filled the valley by 6.0 Ma, that being the age of the uppermost flow. You will see related lava flows to the south and east from here as the geology tour proceeds. It was only after the lava flows filled the ancestral valley that the Oak Creek fault became active and Oak Creek Canyon was carved. Through hundreds of ancient earthquakes the eastern side of the fault dropped about 1,000 feet. Oak Creek Canyon can be no older than 6.0 Ma because the uppermost lava flow is broken by the fault and the canyon could not have been cut before the fault. This makes Oak Creek Canyon about the same age as the Grand Canyon and most other southwestern canyons.

Some may wonder at what rate Oak Creek Canyon was carved and surprisingly, there is some information here. Walk a short distance to the east on the road that brought you up here and you can see some of this evidence. Watch for approaching vehicles as you carefully cross the cattle guard-you are about to look at something that virtually no one else notices. Look for a dark soil in the roadcut and look for the large basalt cobbles within it. Notice that they are rounded smooth, a texture that tells geologists that they were rolled in a streambed. But here they are on top of Airport Mesa, 700 feet above the nearest stream. What could this mean? Airport Mesa (Table Top Mountain on topographic maps) became a great place to land airplanes because it's flat, and free of obstacles. However, it is not completely flat and in fact, has a gradient of about 80 feet per mile down to the southwest. The deposit (about 100 feet thick) broadens to the southwest away from the mouth of the modern canyon. Curiously, modern Oak Creek has a gradient of about 70 feet per mile to the south-west and contains near identical cobbles of rounded basalt. Geologists interpret this to mean that Oak Creek deposited the cobbles here before it cut down 700 feet to its present position on the south side of Airport Mesa. So far, no fossils or other datable materials have been found in the deposit but it must be younger than the Verde Formation, about 2.5 Ma. If we assume an age of 2.0 Ma for this deposit, then Oak Creek may be cutting its canyon at an average rate of about 350 feet per one million years. If this rate were consistent through the last six million years, we would expect the canyon to be 2,000 feet deep, but it is actually closer to about 2,500 feet at its mouth. The discrepancy is easily accounted for when one assumes that the dissection rate has not been constant through time and was probably more active when canyon cutting was first initiated. Detective work at its best from an inconspicuous deposit of perched cobbles!

Return to your vehicle and drive down the road to the stop sign on Highway 89A. Notice Coffee Pot Rock on the skyline while you are at the stop sign. Do not forget to reset your odometer to 0.0 here. Turning left onto the highway is sometimes an exercise in patience. Remember the lessons of geologic time and take it slow.

0.0   Turn left onto Highway 89A. For the next two miles you will travel through West Sedona, known as Grasshopper Flat before this area was urbanized.
0.4   Capitol Butte and Coffee Pot Rock are visible on the right skyline.
1.1   Capitol Butte, resembling a capitol dome, is visible on the right skyline
2.0   Traffic light at Dry Creek Road.
2.3   The Hermit Formation is visible in the small roadcuts on the left for the next 0.5 miles.
2.7   View of Scheurman Mountain ahead, capped with a lava flow that, like House

Mountain, belongs to the Hickey Formation.

3.0   Traffic light at Upper Red Rock Loop Road, the access for Red Rock Crossing at Crescent Moon Ranch.

3.1   Hermit Formation on the right.

3.5   Begin a downhill stretch through the Supai Group on the up-thrust side of the Cathedral Rock fault. Going slow, you may notice on the right a small, vertical **dike** (gray) rising through the Supai Group (red) most likely belong to the Hickey Formation.

3.8   Cross the Cathedral Rock fault with about 220 feet of offset evident here.

3.9   The pyramid-shaped feature ahead is locally called "Rose's volcano." It is not a volcano but may be a vent-concealing flow that was erupted along the Cathedral Rock fault. It sits on the down-thrown side of the fault.

A small volcanic dike cutting through the Supai Group

4.3   Lower Red Rock Loop Rood. It is three miles from here to the entrance of Red Rock State Park, where there are geology exhibits and nature trails. Ranger-led geology hikes are given by knowledgeable park volunteers.

4.4   The Schnebly Hill Formation is exposed on the left for the next 0.5 miles.

4.5   Drive parallel to the Cathedral Rock fault on the right for the next mile. You will see the up-thrust side of the fault in the light red beds of the Supai Group.

4.7   Milepost 368.

5.0   Another view of the pyramid-shaped "Rose's volcano" on the right.

5.4   The House Mountain volcano is visible on the left skyline.

5.7   Milepost 367.

5.8   The Hermit Formation is exposed in roadcuts on both sides of the road for 0.3 miles.

6.2   Bridge across Dry Creek.

6.5   Slowly begin to pull over onto the right shoulder for a short roadside stop at the roadcut ahead.

6.6   Just beyond milepost 366 is a roadcut with a small, well-exposed fault.

Stop #2–Discussion about Faults in the Verde Valley

After parking well off the road, walk over to the roadcut and notice an obvious fault. On the down-thrown (southwest) side, dark gray basalt of the Hickey Formation is in contact with the red Hermit Formation (on the

Fault exposed in a roadcut west of the Dry Creek Bridge. The left (southwest) side is down-thrown.

northeast side). A small remnant of the light pink Beavertail Butte formation is visible on top of the Hermit Formation. About four feet of unfaulted **colluvium** overlies the fault and **bedrock** here. This is one of several faults that down-drop the Verde Valley relative to the Sedona area; these other faults are largely concealed beneath surface rocks as you drive to the south. Collectively, the faults mark the northeast limit of the Verde graben, which lies between the Black Hills to the southwest and the Sedona area. The Verde graben formed after 10 Ma as the Black Hills were uplifted relative to the Verde Valley. It was within this down-faulted **graben** that the Verde Lake was formed between 9.0 and 2.5 Ma. You will see much evidence for this ancient lake in the Verde Formation later on the geology tour.

Faults like the one exposed here may delineate a precise boundary between the Colorado Plateau Province and Arizona's Transition Zone. The erosional edge of the Mogollon Rim serves as a more visible boundary but these faults are perhaps a more realistic one. Return to your vehicle and resume road log.

6.6   As you carefully pull out onto the highway look at the roadcut to the left and notice the massive exposure of basalt rock that progressively gives way to very coarse deposits that contain large basalt **clasts**. This is where the Verde Lake washed over a resistant knob of basalt. The coarse **lithology** belongs to the Verde Formation and you will see many examples of this relationship in the road log ahead.

7.1   The Black Hills are in view on the skyline ahead and to the right. Locally, the high, flat top is called Mingus Mountain and the lavas there belong to the Hickey Formation.

7.6   The wetlands on both sides of the road are where treated waste water from the city of Sedona recharges the water table.

7.7   Milepost 365.

7.8   Sign "4,000 feet" elevation.

8.5   The low, rounded hills ahead are composed of the volcanic Hickey Formation.

8.7   Milepost 364.

9.2   Roadcuts in volcanic rock for the next 0.2 miles.

9.7   Milepost 363. Prepare to turn left ahead by moving to the left lane.

9.9   Exit to the left on the Page Springs Road, Yavapai County Road #50.

10.0   Turn left to Page Springs. Watch for oncoming traffic and after making the turn pull over on the gravel parking area on the immediate right. Get out to view the profile of the House Mountain volcano with a short discussion.

Stop #3–Discussion of the House Mountain Volcano

Visible to the east on the skyline is the broad, gentle profile of the House Mountain volcano. This landscape feature is often overlooked among the colorful and wide-spread red rocks of the Sedona area. But House Mountain tells an interesting story of volcanism that erupted at the base of the Mogollon Rim. House Mountain is just one of many volcanic centers that belong to the Hickey volcanic field and related deposits called the Hickey Formation. The lavas on Mingus Mountain, as well as those between Camp Verde and the Carefree Highway on Interstate 17, are also part of this large volcanic field, which was active between 15 and 10 Ma.

House Mountain is a shield volcano with broad, gentle slopes resulting from the very fluid lavas that flowed for large distances. Its shield profile is obvious from this point of view. Two radiometric dates were obtained from House Mountain and yielded ages of 14.5 and 13.2 Ma. At the time that these dates were obtained, the volcano was erroneously linked with a flow on its south side that was one-third this age. Another curiosity about House Mountain is that the lavas are found only on three sides of the central vent. To learn more about why this is true, please refer to the hike description on page 134. Return to your vehicle, reset your odometer to 0.0, and proceed with the road log.

0.0   The House Mountain volcano is visible on the skyline for the next 0.5 miles.
0.3   Hills with the Verde Formation exposed on the right.
0.5   Roadcut through volcanic rocks from House Mountain.
0.7   Very coarse-grained Verde Formation on the right.

© Wayne Ranney

The House Mountain shield volcano as seen from Page Springs

0.9   Verde Formation on the left.
1.1   Verde Formation in the mesa on the skyline ahead.
1.5   Lolomai Springs is located down the hill to the left; you are now entering an area rich in springs on lower Oak Creek. A discussion of springs in this area will be given at the next stop about 1.5 miles ahead.
1.9   Look for dark volcanic rocks capped with the Verde Formation on the left, next 0.2 miles. This is the result of limestone deposition from the Verde lake that once buried the House Mountain volcano.
2.1   House Mountain basalt and Verde Formation seen ahead. Entering the flood-plain of Oak Creek.
2.3   Pass Bubbling Ponds, a warm water spring, on the right.
2.6   Bridge across Oak Creek.

2.7   Recent **alluvium** on the left.

3.0   Page Springs Fish Hatchery on the right. Pull into the driveway here for a discussion of springs in the Verde Valley. There are rest rooms here and interesting displays related to the hatchery.

Stop #4—Discussion about the Geology of Fresh Water Springs in the Verde Valley
What a wonderful and cool desert oasis this is! Pure, crystal-clear water emerges from the ground here at the rate of about 15 million gallons per day or 10,000 gallons per minute. While it may seem odd at first to see so much water rushing from the ground, it is not that unusual below the Mogollon Rim. That is because the Coconino Plateau above Sedona serves as a catchment area for some of the 18 to 22 inches of precipitation that falls on the plateau surface. Hydrologists believe that about two inches of this total seeps into the ground and travels through the rocks beneath Sedona to feed these springs and many others found below the Mogollon Rim. Page Springs is one of the largest in the state of Arizona. You can see natural spring water emerge from the ground by going up the small hill to the east of the parking area to a chain link fence with the number "7" on it.

Oak Creek (located beyond the parking area to the west) owes its existence to springs. It emerges from Sterling Springs located at the base of the swithchbacks on Highway 89A and is fed by dozens more as it makes its way down the upper reaches of Oak Creek Canyon. Many of these springs are hidden in the creek itself and go unnoticed. By the time the creek flows past Indian Gardens, some ten miles below Sterling Springs, the creek has attained a volume of about 15,700 gallons per minute. (For reference, Sterling Springs, where the creek begins, contributes between 320 and 390 gallons per minute). This stretch is called a gaining reach for a stream since it gains water volume. However, between Indian Gardens and this part of Oak Creek, it is a losing reach with a net loss of surface water absorbed back into the ground. Before reaching Lolomai Springs, Oak Creek holds less than 9,000 gallons per minute. As the creek flows through this reach, it passes by many springs that have colorful names-Lolomai, Bubbling Pond, Tree Root, Turtle Pond, and Page Springs. The creek increases its volume through this gaining reach by about 220 percent to a whopping 28,700 gallons per minute.

Hydrologists have developed sophisticated ways to determine the **residence time** for water in an aquifer (how long the water has been in the ground). Studies here and in other parts of Arizona have shown that the water in some springs has a residence time of hundreds or even many thousands of years. It's incredible to think of water traveling underground for that length of time. However, this does not mean that if water from snowmelt ceased to percolate into the system, that springs in the area would continue to flow for hundreds or thousands of years. The groundwater regime is a complex plumbing system and changes in any part of that system may affect distant parts of that system. Fluctuations in groundwater discharge from springs are not uncommon and have been observed through time. Recent changes in climate and precipitation patterns are a concern and the rapid increase in population here may also alter the groundwater flow from springs. Geologists and hydrologists in the Verde Valley can and will be important players in the groundwater management decisions that must address this important issue. You can take the self-guided tour of the facility here for more information.

Return to your vehicle and proceed to the stop sign. Reset your odometer to 0.0.

0.0   Turn right onto Page Springs Road.

0.1   Coarse deposits of the Verde Formation are exposed in the roadcut on the left for the next 0.2 miles. This material was shed from House Mountain to the shore of the Verde lake.

0.6   After the sharp left turn look ahead on the skyline for low outcrops of the Verde Formation with volcanic rocks from House Mountain.

0.7   Go slow to observe the roadcut on the left containing black basalt interlaced with in-fillings of white calcite. An ancient, cracked surface on the slope of the volcano was slowly infiltrated by the calcium-rich water of the rising Verde lake.

Outcrops like these are common on the old slope of the volcano.

Calcite-filled fractures in the House Mountain basalt

0.9   The Verde Formation is visible ahead on the left.

1.1   Roadcut containing coarse deposits of the Verde Formation. This lithology was deposited near the ancient slope of the House Mountain volcano.

1.2   Coming up on the left is a resistant knob of House Mountain basalt that is blanketed on either side with coarse rubble in the Verde Formation. This knob was a small island as lake water rose around it.

1.3   Pull over on the right before the little brown building to view the roadcut across the way. A resistant knob of basalt is visible on the left side of the cut and as you bring your gaze to the right you will begin to see limestone-cemented rubble that covered this ancient slope. This

Depositional contact of coarse Verde Formation (right) against a remnant of House Mountain basalt (left)

rubble eventually gives way on the right side of the bedded (but coarse) Verde Formation. This is a beautiful cross section of an ancient slope that was preserved by lake deposits.

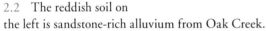

1.7   Coarse Verde Formation in the roadcut on the left.

2.0   Again, there is coarse Verde Formation in the roadcut on the left.

Close-up of the coarse-grained Verde Formation along the Page Springs Road

2.2   The reddish soil on the left is sandstone-rich alluvium from Oak Creek.

2.6   Top of the hill through the Verde Formation.

2.8   Enter the outskirts of Cornville. Mingus Mountain is visible ahead to the right and the town of Jerome can be seen about halfway up the mountain.

4.2   Stop sign at KC Korner. Reset odometer to 0.0 and turn left.

0.0   Yavapai County Road #30.

0.3   A nice view of the House Mountain volcano (with the Mogollon Rim visible to the left of it in the far distance) is on the left for the next 0.3 miles. On the slopes of House Mountain look for the white deposits of the Verde Formation, left as erosional remnants from when the volcano was buried under Verde lake.

The House Mountain volcano from near Cornville. Note the whitish Verde Formation remnants pasted on the side of the volcano.

0.4   Roadcuts are through the Verde Formation.

1.0   The Verde Formation can be seen in the distance on the skyline.

1.5   Pass through eroded remnants and rounded hills of the Verde Formation next 1.5 miles.

2.4   Milepost 7 and the Verde Formation in roadcut on left.

2.9   The Verde Formation is in the roadcuts on both sides of the road. Note how the formation is much finer-grained here than on the Page Springs Road. These deposits were much farther from the House Mountain volcano and the coarser debris deposited on its flanks.

3.1   The round-topped Hackberry Mountain volcano is visible straight ahead in the far distance. It is composed of a volcanic rock called dacite and was active between

10.0 and 7.0 Ma.

3.4   At milepost 8, prepare to turn left on Beaverhead Flat Road

3.5   Turn left onto Beaverhead Flat Road. (Yavapai County Road #78).

3.6   Munds Mountain ahead and Beaverhead Flat Trailhead parking on the left. Rounded hills are cut into the Verde Formation nearby.

Typical exposure of the Verde Formation in the central part of the Verde Valley

3.8   Going downhill, notice Beaver Creek in the distance, where it cuts into basalt from the plateau, which dates between 8.0 and 6.0 Ma.

4.1   Verde Formation on the left. About 0.25 miles north from here is a 5.5 Ma lava flow that was originally was thought to belong to House Mountain. It is interbedded with the Verde Formation and gave an erroneous age for the volcano. Instead the flow originated on the plateau to the northeast and is not from House Mountain.

4.3   View of Munds Mountain and Jack's Canyon straight ahead. The Oak Creek fault trends towards the viewer in Jack's Canyon.

4.5   Milepost 1.

5.0   View of the flat-capped Verde Formation in the hills on the left horizon.

5.5   Milepost 2.

5.7   Small mesa on the right is composed of the Verde Formation.

6.2   Forest Service Road 9952 on the left.

6.5   Milepost 3. Note the thin plateau basalt remnant capping the Verde Formation ahead and on the right.

7.2   Outcrops of the Verde Formation in the roadcut.

7.5   Milepost 4.

8.0   USFS Road 120A on left. This is very rough 4x4 road to the top of House Mountain.

8.5   Milepost 5.

8.6   Turn left on USFS Road 9500G and park your vehicle for a view of the plateau basalt, the Beavertail Butte formation, and a short geology discussion of each.

Stop #5–Geology Discussion of Nearby Features

Park your vehicle and walk between the two waste bins towards the red rock monument to the east. Here you will read about the historic Chavez Trail that once connected the territorial capitals of New Mexico and Arizona. Geologically there is much to see here. First, turn your attention to the unnamed, round-topped butte to your left (north). Notice the thin cap of basalt that forms its rounded top. The basalt cap is a remnant of a lava flow that originated on the plateau to the northeast of here. If you turn your gaze more eastward you will see thicker sections of these same flows

© Wayne Ranney

across Dry Beaver Creek. Geologists once referred to these flows as the "ramp basalt" since they form a ramp-like surface from the plateau edge to the floor of the Verde Valley. More recently, they have been called "sheet flows" since their low viscosity allowed them to spread outward in a thin, sheet-like manner. In this book they are called the "plateau basalt" and they are the same lavas that you saw at Stop #1, which filled an ancestral valley before modern Oak Creek was cut. This is very close to where the ancestral valley opened out into the larger Verde Valley.

The remnant plateau basalt seen on the butte to the north covers volcanic rocks that erupted locally from House Mountain. This older House Mountain basalt is not readily visible in the round-topped butte but is present beneath the grass-covered slope. Other House Mountain basalt remnants are visible to the southeast from

The view east from Stop #5 showing the dissected plateau basalt (mesa tops) capping the Schnebly Hill Formation (mesa slopes). The Beavertail Butte formation forms the valley floor.

here in Beavertail Butte, the hill with a hummocky surface and a winding road to its top. A flow from here was dated at 15.4 Ma. The older basalt rests on gravels that are pinkish in color and can be seen indistinctly in outcrop or the open gravel pit below the viewpoint. These gravels are informally called the Beavertail Butte formation and this area is the type section (or the place where the formation was first described). Note that these outcrops rest on the relatively low Hermit or Schnebly Hill formations and that a higher Paleozoic section exists nearby to the northeast. Geologists interpret this to mean that the gravel here is near its northeast limit, being deposited against a rising wall of upper Paleozoic strata-the ancestral Mogollon Rim.

A summary of the geologic events recorded here is: 1) drainage from the Mogollon Highlands entered the area from the south and west between 25 and 15 Ma and these deposits were deflected to the southeast as they encountered the Mogollon Rim; 2) the deposits were partially eroded when lava from the House Mountain volcano covered them about 15.4 Ma; 3) this lava was in turn partially eroded when the plateau basalt flowed over it between 8.0 and 6.0 Ma; and 4) dissection proceeded to expose all of the deposits and carve the landscape before you. Without the most recent period of erosion, all of this would still be in the subsurface and the ancient history would remain unknown.

Return to your vehicle and the stop sign on the road. Reset your odometer to 0.0 and turn left.

0.1 Go through the roadcuts in the Beavertail Butte formation. Just beyond this is the entrance to the gravel pit seen from stop #5. Visible on the left is a pyramid-shaped outcrop of the Beavertail Butte formation.

0.2   Roadcuts in the Beavertail Butte formation on the left and right for the next 0.5 miles.

1.0   Junction with State Highway 179. Before the stop sign, pull over to the right on the gravel pad. Note the outcrop to your left on the northwest corner of the junction. Capping this outcrop is the Rancho Rojo Member of the Schnebly Hill Formation (bright orange). This unit was deposited as small coastal dunes along an arid coastline about 282 Ma. It rests on the darker-colored Hermit Formation, deposited in ephemeral fluvial conditions.

A typical exposure of the Beavertail Butte formation (reddish pyramid) from near Stop #5

Now look straight ahead across Highway 179 and the channel of Dry Beaver Creek to a small graben that is visible on the far wall. You will note the small displacements within the bright orange Rancho Rojo Member but the faults also offset the overlying Beavertail Butte formation. Like most faults in the area, it shows down-to-the-southwest displacement. Approach the stop sign and reset your odometer to 0.0.

0.0   Turn left on State Highway 179 and see the sign for "Red Rock Country."

0.1   In the low grass-covered hills to the left is a last look at the Beavertail Butte formation. The red bedrock on the left side of the road marks the location of the Mogollon Rim between 25 and 20 Ma. At this spot, the Beavertail Butte formation pinches out to the northeast against a steeply rising slope of Paleozoic bedrock. The ancestral rim prevented the gravel from being deposited any farther north.

0.2   For the next 0.5 miles, on the left the low hills are capped by the thin, bright orange Rancho Rojo Member of the Schnebly Hill Formation. Below the Rancho Rojo Member is the darker Hermit Formation.

0.6   A view of Horse Mesa straight ahead. It is capped by a 6.4 Ma lava flow that flowed off of the plateau into the Verde basin. Interstate 17 to the east was constructed on top of this same basalt surface.

0.9   Sign "Elevation 4,000 feet." Note: Topographic maps show the 4,000-foot contour about 0.5 miles to the north. The Hermit Formation is exposed in the roadcut to the left. Note the channels cut into some beds and filled with other sediment, documenting the shifting nature of the streams that existed in this area about 285 Ma.

1.0   The Hermit Formation is exposed in the roadcut.

1.1   The plateau basalt is visible on the skyline to the right. Note how Rattlesnake Canyon, part of the Dry Beaver Creek drainage, is cut into the basalt plateau.

1.2   The Hermit Formation is visible in the roadcut to the left.

1.4   Ahead is a glimpse of Bell Rock, Courthouse Butte, and Lee Mountain (left to right). Lee Mountain is the southern part of Munds Mountain, which makes up the Mogollon Rim in this area.

1.5   Approaching the USFS Visitor Contact Center on the right.

1.6   Turn right here for a geology discussion, local information, a bookstore, and rest rooms.

Stop # 6–Discussion about the Hermit Formation and Scenery at the Red Rock Visitor Center

The USFS visitor center and Red Rock Ranger District office was opened in 2008 and houses a great geology display, a nonprofit bookstore, and an information center staffed with helpful volunteers from the organization Friends of the Forest. After visiting the geology display inside, look to the north beyond the green railing. You can see Bell Rock, Courthouse Butte, and Lee Mountain

Horse Mesa from the USFS Red Rock visitor center parking lot

from here. More will be said about these landforms at the next stop. To your right is Horse Mesa capped by the plateau basalt dated at 6.4 Ma. Note that the plateau basalt rests near the contact of the red Schnebly Hill Formation with the white Coconino Sandstone. Similar basalt rests on the same contact in Oak Creek Canyon farther to the north. The basalt filled an ancestral valley that was carved down to this stratigraphic level before modern Oak Creek Canyon was cut. Notice that Woods Canyon, to your right, bisects the plateau basalt and separates Horse Mesa on the north from Beaverhead to the south.

Now walk back to the parking lot and the wall that has been cut into the rock to its the right-hand (west) side. This excavation exposes a fresh view of the normally rubble-strewn

Roadcut with multiple stream channels in the Hermit Formation; USFS Red Rock visitor center parking lot

and inconspicuous Hermit Formation. Walk all the way to the south end of the cut where the paved road begins to curve. In this corner exposure you will notice a few cut-and-fill features where ephemeral streams carved small channels that were refilled during floods with silt- to pebble-sized material. This type of bedding is ubiquitous in the Hermit Formation but because it erodes easily it is rarely seen. Fresh cuts like this reveal the secret of the Hermit's depositional history.

Now move slowly back along the wall towards the visitor center and look for grayish zones that extend downward through the red sediment. Look closely and you will see that these features are often filled with calcium- or silica-rich centers that branch and curve downward.

These are root casts, which are the remains of the roots of ancient plants and shrubs that grew on the Hermit floodplain. The organic matter in the roots created reducing conditions (the opposite of oxidized) in the soil and these conditions allowed other mineral material to replace the organic matter in the root. There are numerous examples of root casts in the cut here. It's quite amazing to think that such fleeting features as plant roots or seasonal floods could be preserved through 285 million years of time. Yet here we see the evidence laid bare by the blade of a bulldozer.

Return to your vehicle and continue the road log.

A fossilized root cast in the Hermit Formation; USFS Red Rock visitor center parking lot

0.0   At the stop sign on Highway 179, reset your odometer to 0.0 and turn right.

0.1   In the roadcut to the left, look for the contact of the Hermit and Schnebly Hill formations about six feet above road level.

0.2   Milepost 305.

0.3   Ahead on the left is a great view of a fault with about 30 feet of offset evident. The bright orange Rancho Rojo Member of the Schnebly Hill Formation has been down-faulted to the south against the brick-red Hermit Formation, exposed in the up-thrown side of the fault. After exiting the roadcut there is another good view ahead of Courthouse Butte.

0.5   Ahead and to the right is the Munds Mountain

Exposed fault along Highway 179 with the orange Schnebly Hill Formation (left) down-thrown against the brick-red Hermit Formation (right)

section of the Mogollon Rim.

0.7   Sign "Entering the Village of Oak Creek."

0.8   Passing Horse Mesa on the right.

1.0   In the roadcut to the left is the Hermit Formation with visible root casts.

1.2   Highway roundabout at Ridge Trail Drive. Milepost 306 is just beyond the roundabout and the Rancho Rojo Member will be exposed in the roadcut to the right. You will be driving through the Village of Oak Creek for the next one mile.

1.5   Roundabout connecting the Jacks Canyon Road on the right and the Verde Valley School Road on the left. Continue on Highway 179 straight ahead. The scenic Verde Valley School Road provides access to the House Mountain geology hike (see description on page 134).

1.7   Roundabout at Cortez Drive. You will get a view of Bell Rock ahead, type section of the Bell Rock Member of the Schnebly Hill Formation. This sandstone was deposited initially by the wind but was ultimately reworked along the shore of the Pedregosa Sea, hence the rock's flat-bedded appearance. Courthouse Butte is a much larger monolith to the right of it and is capped by the Coconino Sandstone.

2.3   Roundabout at Bell Rock Boulevard; leaving the Village of Oak Creek.

2.4   Prepare to turn right at the Bell Rock Vista and Pathway interpretive area. Pull as far north into the parking area as you can for trailheads and a discussion of the geology here.

Stop #7–Discussion about Bell Rock, the North Side of House Mountain, and Horse Mesa

This is a fantastic area with lots of opportunities for easy hiking. It is the trailhead for the Bell Rock Pathway and other trails past Courthouse Butte and Bell Rock. Here we will discuss the geology of Bell Rock, the significance of the position of the House Mountain volcano to the southwest, and Horse Mesa to the south. Display your Red Rock Pass before leaving your vehicle.

Walk to the north end of the shelter for a wonderful view of Bell Rock, the type section of the Bell Rock Member of the Schnebly Hill Formation. Bell Rock sits atop the Hermit Formation, which is not visible from here but is present just below the floor of the valley. Note the flat-bedded sandstone in Bell Rock, interpreted to have been deposited in **eolian** settings but reworked by sea tides some 280 Ma. The Bell

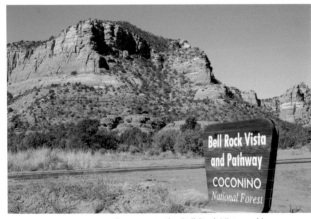

Entrance to the Bell Rock Vista parking area

Rock Member is 500 feet thick here but grades out to the northwest and intertongues with the exclusively eolian Sycamore Pass Member. This relationship shows that eolian conditions were more prevalent to the northwest of here at that time.

Now look closely to the top of Bell Rock and you'll see the Ft. Apache Member near the very top of the spires. The Ft. Apache Member is a 10-foot-thick layer of limestone and dolomite that pinches out just to the west of Sedona but thickens to over 100 feet at its type section in east-central Arizona. It represents a brief inundation by the Pedregosa Sea in the present-day Sedona area about 280

Bell Rock from the south

Ma. This relationship shows that terrestrial environments were more prevalent to the northwest at this time, and that shallow marine conditions existed to the southeast (see paleogeographic map on page 44). Note the position of the Ft. Apache Member on Courthouse Butte to your right (east) and you'll see that it is in the middle of this massive butte at a slightly different elevation. A small fault offsets the strata between these two monoliths with down-to-the-west displacement. The Cathedral Rock fault also offsets the strata between Bell Rock and the informally named Seven Sisters to your left (west). Highway 179 takes advantage of this geologic break as it winds its way towards Uptown Sedona to the north. If you like, you can follow the trail signs north to get a closer view of Bell Rock, or take the trail east on a loop around Courthouse Butte. The Mogollon Rim is visible from here on Lee Mountain. It is the high-standing cliff of upper Paleozoic rocks behind Courthouse Butte to your right (northeast).

Now turn around to face the south and look out over the open valley that was called Big Park before it was urbanized. Slightly to your right on the skyline (south-west) is a long low ridge of basalt that is the northern limit of the House Mountain volcano. Remember its shape on its other side and note that the typical volcano form is not evident from here. This is because of the asymmetrical extent of the House Mountain lava flows, which radiate away from the central vent in all directions except to the north. This is the result of the eruption of lava beneath the Mogollon Rim when it was located in that position. The ridge of basalt you can see to the southwest marks the approximate position of the flows as they were erupted and ponded against the rim some 15 to 13 Ma. The ridge is an example of inverted topography, whereby something that was once low in a valley is now high on a mesa due to differential erosion. The rim has retreated in the last 13 Ma to its present position at Lee Mountain. The distance between the northeast edge of House Mountain and Lee Mountain is about four miles, giving an average rate of retreat of about one foot every 625 years.

Now look to Horse Mesa, the long, flat mesa seen on the skyline to the south and east. The capping basalt is dated at 6.4

View southwest from the Bell Rock Vista parking area. The mesa top is the north edge of the House Mountain volcano and marks the position of the Mogollon Rim about 13 Ma.

Ma and belongs to a set of flows from the north that filled an ancestral valley before Oak Creek Canyon was cut. These are the same flows seen in the east wall of Oak Creek Canyon from the Airport Mesa stop. The Oak Creek fault is present to the east and separates the upper Paleozoic strata on Lee Mountain from the basalt plateau to the east of it. The whitish deposit seen below the basalt on Horse Mesa is similar to other deposits found beneath the flows in Oak Creek Canyon. Most likely, the flows on Horse Mesa also represent inverted topography from 6.4 Ma, when the ancestral valley opened into the larger Verde Valley to the south. The distance from the flows on Horse Mesa to Lee Mountain is about two miles, giving an average rate of retreat of about one foot every 606 years, nearly identical to that documented from the House Mountain flows. Return to your vehicle and continue with the last part of the road log.

0.0   Return to the stop sign on Highway 179 and reset your odometer to 0.0. Turn right.

0.2   Passing Courthouse Butte and Lee Mountain on your right.

0.5   Now passing Bell Rock on the right. Notice the flat-bedded sandstones, which have developed into a classic slickrock terrain. The roadway is constructed above the trace of the Cathedral Rock fault, which has lowered the left-hand (west) side over 300 feet relative to Bell Rock.

0.6   The Seven Sisters (informal local name only) on the left capped by Coconino Sandstone.

0.8   Milepost 308.

0.9   County line between Coconino and Yavapai counties. Drive past the scenic vista coming up on the right.

1.2   View of Twin Butte ahead. This divided highway was constructed in 2008.

1.8   Milepost 309 with a view of Capitol Butte ahead, resembling a capitol dome.

2.0   View of Cathedral Rock ahead and to the left, composed entirely of the Schnebly Hill Formation. The floor in the prominent gap in the middle of this monolith is held up by the Ft. Apache Member.

2.1   Sign "Sedona City Limit."

2.7   Sign "Entering Sedona" and the end of the officially designated Red Rock Scenic Road.

2.8   Milepost 310.

2.9   Roundabout with Indian Cliff Drive on the right and the Back O' Beyond Road to the left. The Back O' Beyond Road provides access to the geology hike on Cathedral Rock on page 138.

3.2 View of Wilson Mountain and Wilson Bench straight ahead for 0.2 miles. A

Cathedral Rock from the east along Highway 179

branch of the Oak Creek fault separates these two; the bench was down-thrown about 860 feet relative to Wilson Mountain.

3.3 Roundabout with Chapel Road to the right.

3.5 The flat-topped feature ahead is Airport Mesa or Table Top Mountain, capped by a 100-foot-thick section of sediment that is identical to what can be found in the bed of Oak Creek today. This suggests that Oak Creek deposited this material before the creek incised 700 feet to its present depth. No fossils or other datable material has been found but down-cutting rates from other nearby areas suggest this has occurred in the last 2.0 Ma.

3.8 The Hermit Formation is in the roadcut on the right.

4.3 Munds Mountain is on the skyline ahead for the next 0.2 miles and exposes the entire upper Paleozoic section of strata.

4.6 The Hermit Formation is exposed in the left roadcut.

4.7 Wilson Mountain and Wilson Bench visible ahead on the skyline for the next 0.2 miles.

Faulted lava flows on top of Wilson Mountain (left) and Wilson Bench (right). The amount of offset is about 830 feet.

5.0   Descend an obvious hill and cross a wash with Hermit Formation exposed nearby.

5.8   Schnebly Hill Road joins on the right and goes to the top of the basalt plateau and eventually on to Interstate 17. (This dirt road is suitable for all vehicles but is definitely not a highway and is very bumpy.) Cross the bridge over Oak Creek, which was constructed in 2009. Just past the bridge is the entrance to Tlaquepaque shopping area.

6.3   Roundabout at the "Y" junction of Highways 179 and 89A. You have now completed the road log and deserve a cold beverage in one of uptown Sedona's nearby establishments.

GEOLOGY ROAD LOG 2
Highway 89A through Oak Creek Canyon

This is a 16-mile road tour that will take you from the heart of Uptown Sedona into Oak Creek Canyon and up the famous switchbacks to the top of the Mogollon Rim. It will pass Midgley Bridge, the Oak Creek fault, Sliderock, the West Fork area (Call of the Canyon), and end at the Oak Creek Vista.

0.0   Set your odometer to 0.0 as you go through the roundabout at the junction of State Highway 179 and Highway 89A (formerly the "Y"). Travel north on Highway 89A as you enter the urban area of Uptown Sedona for 0.6 miles.

0.6   Cliffs composed of the Supai Group are visible to the right (east) over the building tops and across Oak Creek.

View from Uptown Sedona of the Supai Group (lower cliff), the Hermit Formation (tree-covered slope) and the Schnebly Hill Formation (upper cliffs)

© Wayne Ranney

On the southern side of the cliffs, the Supai Group is set against the younger Hermit Formation and marks the east/west trace of the Sedona fault. Cross the Sedona fault with as much as 400 feet of offset down to the southwest. The uplifted rocks of the Supai Group (Esplanade Sandstone) are exposed in the high roadcut on the left.

0.8   Milepost 375.

0.8   Road begins to climb out of the Supai Group exposed in the roadcut to the left and in the canyon wall on the right.

1.3 Begin Oak Creek Canyon Scenic Road and the Oak Creek Canyon Recreation Area in the Coconino National Forest.

1.5 The Hermit Formation is now at road level on the left with the Supai Group in the canyon below on the right.

1.7 Look for milepost 376 before crossing Midgley Bridge. Use your left turn signal well in advance of the small (and often crowded) parking area immediately past the bridge. There is a great view and a worthwhile discussion of the geology exposed here. Display your Red Rock Pass before leaving your vehicle.

Stop #1–Geology Discussion from Midgley Bridge

Walk north past the fee kiosk and the shelter with two picnic tables on a road that predates Midgley Bridge. This road went around the head of Wilson Canyon before the bridge was constructed in 1939. Walk to the first power line where you will have a great view of Wilson Mountain on the left (west) and Wilson Bench on the right (east). Note the difference in elevation between the two landforms, which is the result of offset on a splay of the Oak Creek fault. Both features are capped by lava flows that range in age from 8.0 to 6.0 Ma. Notice how the flows on Wilson Mountain pinch out a short distance to the west and are set upon a rising slope cut into upper Paleozoic rocks. Those on Wilson Bench rest on an eroded surface cut across the top of the Schnebly Hill Formation. Geologists interpret this to mean that the lavas filled an ancestral valley that was cut into the edge of the Mogollon Rim before 8.0 Ma. The lavas and ancestral valley thus predate movement on

The Oak Creek fault (saddle) from near Midgley Bridge

the fault and subsequent canyon cutting. The age of the upper lava flow constrains the faulting and canyon cutting to a time after 6.0 Ma. The amount of displacement between Wilson Mountain and Wilson Bench is about 830 feet with down-to-the-east normal movement. A second strand of the fault is located farther to the east of Wilson Bench along Oak Creek Canyon. This eastern splay has an additional down-to-the-east displacement of about 160 feet, meaning that the total displacement on the whole system is about 1,000 feet. There is no evidence that Oak Creek fault has moved in historic time.

As the main branch of the Oak Creek fault passes between Wilson Mountain and Wilson Bench towards the viewer, it turns to the right (east) sharply at 90 degrees (the smaller eastern splay terminates against this east/west trend). After heading approximately one mile to the east, the fault turns sharply again to the south. It may be that the short east-west segment acts as a "connector" fault between two **en-echelon** north/south faults. The Oak Creek fault continues south on the east side of Munds Mountain where it disappears beneath lavas on Horse Mesa near the Village of Oak Creek. It trends to the north within Oak Creek Canyon and disappears beneath lavas associated with the San Francisco Volcanic Field near Flagstaff. Some geologists have speculated that the San Francisco Peaks stratovolcano may have erupted along a line of crustal weakness related to the Oak Creek fault. Before leaving this area, look up to the west (below Wilson Mountain) to Steamboat Rock and an obvious shoulder beneath its crest. Resistant limestone of the Ft. Apache Member of the Schnebly Hill Formation forms this planar surface. Wilson Canyon, to your left, is cut into the Esplanade Sandstone of the Supai Group.

Now walk back to the parking area and look for the sign to a vista of the Huckaby Trail. Go down the steep set of red-rock steps, beneath the bridge, and to the obvious viewpoint above Oak Creek. Direct your attention to Oak Creek below you, which did not cut its canyon along the trace of the fault here. Rather, the creek leaves the fault zone just upstream from here and turns to the southwest where it flows through unfaulted Paleozoic rocks. This geologic occurrence might explain why so many visitors tend to stop in this area for pictures-the canyon upstream is cut through relatively drab-colored basalt rocks exposed in its eastern wall, while the canyon here is enclosed entirely by colorful upper Paleozoic strata on both sides. Why does Oak Creek choose to not follow the fault in this area, after following it for 14 miles? This is an interesting question that may require an innovative answer.

Perhaps the course of Oak Creek was not determined, as is often perceived, by water finding its way in the downstream direction. Instead, the stream's course could have been placed through geologic processes known as sapping and headward erosion. To

Scenery from Huckaby Trail Vista beneath Midgley Bridge

understand how this works we must first envision the setting of this area before the canyon was carved. Within this pre-canyon landscape, groundwater to the north of here was directed in the subsurface along the trace of the Oak Creek fault from north to south. However, as this groundwater approached the 90-degree bend in the fault near Midgley Bridge, it may have found an easier path to the south by percolating through unfaulted but porous sandstone. This flow would have emerged as springs into a "lower" Oak Creek drainage, with its ancient headwall in the vicinity of uptown Sedona. When groundwater leaks to the surface in springs it tends to undercut the overlying rocks and this is known as sapping. Through time, as sapping progressively collapses more rocks, a stream's channel can be lengthened in the upstream direction and this is known as headward erosion. Thus, a lower Oak Creek drainage may have chiseled its way slowly to the north, carving its course through a headwall of unfaulted red rocks eventually intersecting the Oak Creek fault in the vicinity of Midgley Bridge. Continued groundwater sapping and headward erosion then directed incision along the fault to the north to give us the specific placement we see today. This hypothesis has not been tested by other geologists but is a good explanation for why Oak Creek does not follow the fault south of Midgley Bridge.

Before returning to your vehicle look down the V-shaped gorge of lower Oak Creek Canyon to the southwest for a view of uptown Sedona, Table Top Mountain or Airport Mesa, the subdued profile of the House Mountain volcano, and a partial view of the Black Hills (Mingus Mountain) on the far skyline. You are about to leave this open area and enter the narrow confines of Oak Creek Canyon. Once at your vehicle reset the odometer to 0.0 and turn left onto Highway 89A.

0.0   For the next 0.9 miles the roadcuts on the left reveal the Hermit Formation. On your right is Oak Creek Canyon where the light-colored Supai Group makes up the walls of the inner gorge.

0.5   Grasshopper Point area with a great view to the north of Oak Creek Canyon.

View southwest from the Huckaby Trail Vista. Visible from here are the Supai Group rocks (right canyon wall), Airport Mesa (pyramid-shaped rock in upper left), and House Mountain volcano (far distance).

The eastern splay of the fault is located where the canyon has been cut.

0.8 Milepost 377.

1.2 Crossing the east-west trend of the Oak Creek fault and passing from the up-thrust to the down-thrown side. Note that the Supai Group rocks are tilted skyward in a manner that displays **contrary bedding**-that is, the rocks are bent in a way that appears contrary to the east-side-down movement of the fault. This is evidence of an earlier period of movement on the fault that was opposite (or contrary to) the more recent west-side-up movement. This earlier phase of faulting shoved the east side of the fault higher than the west side and is related to a period of compression known as the **Laramide Orogeny**. The ancestral valley that was mentioned in the discussion at Midgley Bridge may have been cut when erosion attacked the uplifted strata on the eastern side of the fault.

1.6 Schnebly Hill Formation in roadcut to the left next 0.2 miles.

2.1 Historic Indian Gardens where Apache Indians were growing beans and squash when the first settlers came into the area. A flood in February, 1993 covered the highway at this spot.

2.5 Schnebly Hill Formation in roadcut on the left and in other cuts for the next 2.0 miles.

2.7 The plateau basalt is exposed on your right, high in the east wall of Oak Creek Canyon. This is lava that filled an ancestral valley between 8.0 and 6.0 Ma and was later down-thrown by movement on the Oak Creek fault.

2.9 Colluvium on the left eventually gives way to exposures of the Schnebly Hill Formation.

3.0 Milepost 379.

3.1 Cliffs on the left are composed of the Schnebly Hill Formation. Note the magnificent cross-beds exposed in many of the horizons.

3.4 Encinoso Picnic Ground on the left and a trail to Wilson Bench and the top of Wilson Mountain (Forest Trail #123). You will be able to see the valley-filling basalt resting on top of the Schnebly Hill Formation in the high cliff ahead. During spring snowmelt or summer rains, this cliff has a spectacular waterfall.

3.7 Schnebly Hill Formation on the left for the next 0.3 miles.

4.0 Milepost 380. Sign "Elevation 5,000 feet"; many oak trees in this area giving the canyon its name.

4.2 Passing a vegetation-covered debris flow on the left.

4.5 A quick look on the left at the whitish Ft. Apache Member in a roadcut.

5.0 Milepost 281.

5.1 Slide Rock State Park on the left (nominal entrance fee). This is a popular area where a natural slide in Oak Creek slices through the Schnebly Hill Formation.

5.2 The Oak Creek Bridge at Slide Rock. This bridge replaced an older one in 1992. Beyond the bridge and to the right is an outcrop of the Schnebly Hill Formation. There is no parking along the roadway for the next 1.0 mile.

5.6 A good view of the western wall of Oak Creek Canyon for the next 0.4 miles. The entire upper section of the Paleozoic rocks is exposed, as well as a small remnant of basalt. This thin basalt represents the distal edge of the lavas that filled the ancestral valley between 8.0 and 6.0 Ma.

5.8   Fault gauge (whitish powdered rock) associated with the Oak Creek fault on the right. You have crossed the fault inconspicuously many times along the road and will do so ahead as well.

5.9   Use this turnout if you are a slow, observant geologist.

6.0   Milepost 382.

6.3   Historic Garland's Oak Creek Lodge is located on the left and as you go around the curve note the wall on the right meant to hold back runoff debris and faced with artificial "sandstone cross-beds."

6.7   Go slow to view a small dike exposed on the right within the Coconino Sandstone. This dike was probably feeding the plateau basalt flows that filled an ancestral valley before the modern canyon was cut.

6.8   The Coconino Sandstone is exposed on the right, for the next 0.6 miles.

7.0   Milepost 383.

7.8   View ahead to the high cliff in the west wall of the canyon to the Coconino, Toroweap, and Kaibab formations in ascending order.

7.8   Cross Hoel's Wash.

8.0   Milepost 384.

8.4   View to the left of the conflu-ence of the West Fork with the main branch of Oak Creek. The upper part of the Schnebly Hill Formation is exposed at water level. Prepare to park 0.3 miles ahead at Call of the Canyon.

8.7   Signal left well in advance of the Call of the Canyon day use area. A fee is charged to park here and is well worth it if you want to hike in the West Fork area. If you do not have time to hike now, reset your odometer to 0.0 as you drive by or, after turning in, make a U-turn before the fee station.

Winter scene in Oak Creek Canyon

© Wayne Ranney

Stop #2–Call of the Canyon, West Fork, and Discussion about Upper Paleozoic Stratigraphy

You can observe the **stratigraphy** easily from the parking lot or walk 0.25 miles downstream for a slightly better view in the mouth of West Fork Canyon (West Fork Trail, Forest Trail #108).

The West Fork of Oak Creek is probably the most beautiful side canyon in all of Oak Creek Canyon and requires crossing the creek numerous times. The area received its moniker, Call of the Canyon, when Zane Grey wrote his novel of the same name while staying at the Mayhew Lodge, the ruins of which are present near the mouth of the West Fork. The colorful rocks here record the environmental conditions that were present on the far western edge of Pangaea between 280 and 270 Ma (see the paleogeographic map on page 64 and the stratigraphic column on page 17).

The Kaibab Formation is the youngest Paleozoic rock unit on the southern Colorado Plateau and it caps the Mogollon Rim in the Sedona area (wherever basalt is absent). It was deposited in an open marine setting that became shallower and more restricted towards the east. Limestone is the dominant lithology with lesser amounts of chert, gypsum, and sandstone. It is approximately 300 feet thick in Oak Creek Canyon and can be recognized in outcrop as resistant, horizontally-bedded, white limestone that tends to form shallow overhangs.

View of canyon stratigraphy near West Fork. Red strata on right are the Schnebly Hill Formation capped by gold-colored Coconino Sandstone (slope on far left), Toroweap Formation (white cliff) and the Kaibab Formation (upper tree-covered slope).

The Toroweap Formation underlies the Kaibab, but the use of this name in Oak Creek Canyon is complex. Most geologists agree that a Toroweap equivalent is present here and can be differentiated, but what to call it is debatable. The problem arises because of a dramatic **facies** change within Toroweap-age sediments along a north-south strand line from Sedona to Marble Canyon, near the Utah/Arizona state line. This facies change grades from carbonates and **evaporites** in the west to sandstone towards the east. It has been interpreted as a gradation from near-shore marine or **sabkha**-like environments (west) to coastal eolian conditions (east). About five miles east of Oak Creek the Toroweap and Coconino are indistinguishable and are considered a single, eolian deposit.

Dr. Ron Blakey proposed that the Toroweap equivalent in Oak Creek Canyon be named the Cave Springs member of the Coconino Sandstone. This same horizon is present in Canyonlands National Park, Utah, and is called the White Rim Sandstone. Blakey also proposed to name the Harding Springs member of the Coconino Sandstone for the lower part of the formation. I prefer to use the name Toroweap

Formation for a few reasons: 1) stratigraphically, I am a lumper, not a splitter; 2) the facies change within the Toroweap is not quite complete within Oak Creek Canyon and a clearly defined contact is visible between the two formations and known as the "green line"; and 3) other geologists also recognize the facies change within the Toroweap here but prefer to call the rocks an eastern "phase" of the Toroweap Formation. I have absolutely no quarrel with Ron Blakey and his nomenclature scheme-it's just a matter of preference. The Toroweap Formation is about 300 feet thick in Oak Creek Canyon.

The Coconino Sandstone consists of about 600 feet of cross-stratified, pure quartz sandstone deposited in a migrating eolian dune environment. The top contact is sharp, **unconformable**, and easily recognized by a "green line" of vegetation that grows at the contact with the overlying rocks. The "green line" can be seen from the West Fork area by looking to the south towards the middle of the massive cliff above the confluence of the two streams. Note the bench where the pines and manzanita shrubs gain a foothold in the otherwise vertical cliff. The Toroweap is white in color while the Coconino takes on more golden hues. The basal contact of the Coconino Sandstone is gradational and intertongues with the Schnebly Hill Formation, but a contact is generally drawn at the uppermost red sandstone. While looking at the cliff from the Call of the Canyon parking area, note the obvious break in slope below the "green line" that develops on the south-dipping cross-beds of the Coconino Sandstone.

The lowermost unit exposed here is the Schnebly Hill Formation. This is a sequence of mostly red, cross-stratified and ripple-laminated sandstones, with minor but stratigraphically important marine carbonates called the Ft. Apache Member. The Schnebly Hill Formation is not present in the Grand Canyon and was only formally described in 1990. The Schnebly Hill Formation has four members. In descending order they are the Sycamore Pass, the Ft. Apache, the Bell Rock, and the Rancho Rojo members. Generally, the Schnebly Hill Formation records deposition at the margins of the Pedregosa Sea. For a brief period in time, this seaway encroached from the southeast to the site of present-day Sedona, leaving the Ft. Apache Member (still in the subsurface here) as a convenient marker horizon between the Sycamore Pass and Bell Rock members. At the junction of the West Fork and Oak Creek, trough cross-bedding in the Sycamore Pass Member records deposition in eolian environments.

Return to the stop sign at the highway and reset your odometer to 0.0. While at the stop sign, notice the geologically inspired retaining wall directly across the highway, which was built by the Arizona Department of Transportation in 1995. They did a great job of recreating the look of cross-bedded sandstone. Turn left.

0.3   Milepost 385.
0.4   Note the Toroweap Formation in the roadcut to the right, which was deposited along the beach of an arid coastline.
0.6   Note the cross-bedding in the Coconino Sandstone high in cliff ahead.
0.7   Toroweap Formation on the right for the next 0.2 miles.
0.9   Fault gouge on the right.
1.1   Cave Springs Campground on the left.

1.2   The pyramid-shaped monolith ahead was formed when Oak Creek cut its channel to the east of the fault. The road follows the creek and the fault is to the left of the pyramid.

1.3   Milepost 386.

1.5   View ahead displays the gold-colored Coconino Sandstone overlain by the white Toroweap and Kaibab formations.

Constructed retaining wall at the entrance to Call of the Canyon parking area resembling natural cross-bedded sandstone

1.9   Pine Flat Campground on both sides of the road for the next 0.2 miles. Pine Flat is underlain by colluvium washed down from the walls of Oak Creek Canyon.

2.2   Good spring water available on the left. Milepost 387 just beyond this.

2.4   Exposure of colluvium on the right for the next 0.3 miles.

3.0   Crossing the bridge at Pumphouse Wash, the main artery of upper Oak Creek.

3.1   Toroweap Formation on the right for the next 0.2 miles.

3.2   Sign "Elevation 6,000 feet." You have ascended 1,500 feet since leaving Uptown Sedona.

3.3   Toroweap Formation on the left.

3.4   The hairpin turn to the right is necessary to avoid road building through harder rocks in the up-thrown side of the Oak Creek fault. After completing the turn look to the left for a view of the fault gouge formed in the Oak Creek fault zone. In this wide zone of breakage, the rocks have been pulverized to a fine powder and runoff has scoured this zone of weakness to create Oak Creek Canyon.

3.9   Colluvium exposed in the roadcut on the right gives way to an exposure of the Kaibab Formation and more fault gouge.

4.1   A retaining wall built with basalt boulders.

4.2   Hairpin turn to the right is again necessary to avoid cutting through harder rocks in the upthrown western side of the Oak Creek fault. The fault gouge zone becomes visible again on the left.

4.3   A small fault is visible on the left, with black basalt boulder rubble next to white Kaibab Formation debris. The faulted basalt documents that it was emplaced before faulting occurred.

4.5   Thin, red paleosol (ancient soil) on the left at eye level.

4.7   Crude basalt columns exposed in roadcut to the left.

4.9   Great view of Oak Creek Canyon to the right. Parking ahead in another 0.4 miles. The roadcut to the left exposes various basalt flows that are interbedded with thin red paleosols. These colorful contacts represent many thousands of years between the various flows.

5.3   Milepost 390. Prepare to turn right into the Oak Creek Vista parking area.

5.4   Entrance to the Oak Creek Vista parking area at the top of the switchbacks. After parking your vehicle walk along an old alignment of the highway to look down the trace of the Oak Creek fault and canyon. A final geology discussion follows.

A red paleosol trapped between lava flows along the switchbacks, Highway 89A

Stop #3–Oak Creek Canyon Vista and Geologic Summary of the Canyon

Here you are standing on Arizona's second most prominent landform, the Mogollon Rim. The rim forms the boundary between the Colorado Plateau province to the north and the Transition Zone to the south. (Faults located south of the rim may actually define a more sensible geologic boundary but the rim is more easily seen and defined.) Oak Creek Canyon is an erosional **reentrant** into the rim and there are a number of interesting features visible in this vista that shed light on the sequence of events that helped produce this landscape. Walk past the jewelry vendors to the overlook that faces south.

First notice the asymmetrical nature of the canyon. The left-hand (east) wall of the canyon is about 700 feet lower than the west side. This is due to down-to-the-east normal movement on the Oak Creek Fault during the last 6.0 Ma (date obtained from the uppermost basalt). The course of Oak Creek Canyon has been cut, for the most part, directly on top of the fault zone and as you made your way up the canyon you crossed the fault zone several times. The pyramid-shaped features you see rising from the canyon floor are erosional remnants that show where the creek cut a meander bend to the east away from the fault zone.

Next, notice that the west or uplifted block of the canyon wall has a completely different succession of rocks than the down-thrown side. The higher west cliff is composed of upper Paleozoic strata that is white to buff in color and exposes, in descending order, the Kaibab, Toroweap, Coconino, and Schnebly Hill formations. The spectacular red rocks seen downstream near Sedona are not fully exposed in this upper part of the canyon but, along with the Schnebly Hill Formation, they include the Hermit Formation and the Supai Group (again in descending order). The western wall of Oak Creek Canyon is composed almost entirely of upper Paleozoic strata with occasional thin remnants of basalt preserved on top, which thicken towards the canyon.

The east side of the canyon wall is much different than the west side. It is composed of successive layers of dark basalt, which reach a thickness of over 500 feet in the southern part of the canyon. The range of ages for these flows is bracketed between 8.0 and 6.0 Ma. Geologists at Northern Arizona University mapped the canyon and provided a possible sequence of events: Laramide-age movement on the Oak Creek fault resulted in uplift on its eastern side; erosion on this block formed an ancestral valley into the upper Paleozoic layers; this ancestral valley was filled with at least five different lava flows between 8.0 and 6.0 Ma. Faulting became active after lava emplacement but this time with down-to-the-east motion; modern Oak Creek Canyon was carved along the fault after faulting. No movement has been recorded in historic times along the Oak Creek fault.

This is the end of the road log. Continue north to Flagstaff and the Grand Canyon or return to Sedona.

View to the south in Oak Creek Canyon. Note the difference in elevation between left and right. Wilson Mountain on the far skyline.

GEOLOGY HIKE 1
House Mountain Volcano

Although known primarily for its red rocks, Sedona has its very own volcano that lies secluded among thick stands of pinyon and juniper trees southwest of the Village of Oak Creek. Easily overlooked by casual tourists, House Mountain offers an unparalleled story of a time when red-hot lava spilled onto red sandstone. A surprising discovery awaits those who make this moderately strenuous hike to the top of the volcano-a view into its central crater and a fabulous view back to the red rocks. The story gets even more interesting as you near the top.

Access

From Uptown Sedona: From the roundabout connecting highways 89A and 179. Drive south on Highway 179 about 7.8 miles to the roundabout with the Verde Valley School Road. Reset your odometer to 0.0.

From the Village of Oak Creek: Proceed to the roundabout connecting Highway 179 and Verde Valley School Road. Reset your odometer to 0.0.

Approach: Turn west on Verde Valley School Road and continue 3.7 miles on pavement and an additional 0.5 miles on an improved dirt road. At odometer reading 4.2, look for a brown recreation sign that directs you to turn left onto an unimproved dirt road to the Turkey Creek Trail. The final stretch of this road is extremely rough. Many hikers opt to park at this junction and continue on foot. If you have high clearance or lots of experience dodging boulders, you can continue an additional 0.5 miles to a fork in the road. Stay left for another 0.1 miles to a small parking lot. Display your Red Rock Pass before beginning the hike.

A Moderately Energizing Hike to an Extinct Volcano

Look for the steel sign that marks the trailhead on the west side of the parking area (USFS Trail #92). The hike is about 3.2 miles long to the top of the crater and climbs about 800 feet in elevation, most of which comes near the very end. (If you have parked back at the junction with Verde Valley School Road, the total distance is about 3.8 miles one-way.) Remember to look for and follow the caged rock "cairns" whenever you come across other social trails along the way. The hike begins on a very gentle uphill grade on an old jeep road that goes about 1.2 miles to Turkey Creek Tank.)A tank is a constructed depression along a stream that collects water for cattle.) Turkey Creek Tank is oftentimes dry but the large cottonwood trees always give away its location. Another steel sign west of the tank shows you the way to the final part of the hike. This section of trail is steeper with a few switchbacks at the end. You will know you are getting near the top when you start to see red cinders lying on the ground.

House Mountain's Geology in Brief

House Mountain is a relatively subdued landform in the Verde Valley. With its black volcanic rock and gentle profile, it is all but hidden among the stunning red-rock scenery found nearby. But what House Mountain lacks in its visual awe-power is more than made up for with its geologic story. Originally it was thought to be about five and a half million years old, but radiometric dating in the late 1980s proved that it was three times older-between 15 to 13 Ma. This makes it the same age as the lavas above Jerome in the Black Hills (called Mingus Mountain by the locals). If you drove to Sedona from Phoenix, all of the black basalt rock observed between Black Canyon City to Camp Verde are related in age (between 15 and 10 Ma) and belong to the Hickey Formation. It is surprising to some visitors that Arizona contains so much evidence of past volcanic activity.

Ironically, most of this hike is across the red sandstone of the Schnebly Hill Formation. From the parking area to well beyond Turkey Tank, you will be surrounded by buttes and spires of this colorful deposit, laid down in sand dunes that encroached upon the shoreline of the Pedregosa Sea (see the paleogeographic map on page 44 to get a sense of this ancient environment). Many of the hikes in Sedona begin with lots of climbing uphill, but on the start of this hike you can stretch your legs and stride out in rapid fashion. Along the way you might

House Mountain from the southwest

wonder, "If I am climbing a volcano, how come the hike is laid out on sedimentary red rocks most of the way?"

This was the dilemma I ran into, when I set out to make a geologic map of the area for my Master's thesis at Northern Arizona University in the 1980s. Although its profile is subdued compared to most of the red rock cliffs, House Mountain really does look like a volcano when observed from the east, south, or west. This type of feature is called a shield volcano because in profile it looks likes a warriors shield lying on its side. The lava that erupted from House Mountain was quite fluid and flowed readily before cooling to give its subtle but unmistakable shape. For some strange reason no lava flows are found on its north side and it is upon this terrain that you make your approach to the top of the volcano.

There were a couple of possibilities that could explain why no lavas were preserved on the north side of House Mountain. The hands-down favorite when the mapping project began was that the north side of the volcano subsequently had been faulted upwards such that erosion had removed the uplifted lavas. But when I went to the contact where the lava met the red rocks, it was not faulted. Rather, it was a place where lava had erupted onto an already eroded and inclined surface that was cut into the Schnebly Hill and Coconino formations. This observation led to the (initially) outrageous idea that House Mountain had erupted against the side of a cliff face. Sure enough, there is additional evidence to support the hypothesis.

As you make your way past Turkey Tank you will hike up and down a bit but eventually you'll enter a small wash. Keep looking to your left (east) in this stretch and you'll begin to see an area above you where black lava from House Mountain rests atop the red and white sandstone rocks. Note that this contact is not as steep as faults in this area and this was the initial clue that something odd might possibly explain the lack of lava on the volcano's north side. When I finally visited this contact, I saw that it was not faulted at all. Rather the lava had erupted onto a cliff face cut into sedimentary rocks. At that moment, a light brightly illuminated House Mountain's distant past. It is now commonly accepted that House Mountain erupted at the base of the Mogollon Rim before the rim had eroded back to its present location (about four miles away to the northeast on Lee Mountain). What an incredible story!

Contact of the ancestral Mogollon Rim (left) and the House Mountain lava (right)

As you walk in and out of the wash on the uphill climb, you'll begin to notice many different colored rocks on the ground that stand out against the red-rock background: a white limestone here, a gray limestone there, and, if you are observant, a granite-looking rock flecked with black and white crystals. Knowing that rocks always roll downhill, you can impress your friends with your powers of prediction by telling them the kind of rock you'll find at your destination. The bright white limestone is part of the Verde Formation, and the inconspicuous gray limestone belongs to the Ft. Apache Member of the Schnebly Hill Formation. And those curious crystalline rocks formed in the magma chamber that fed the House Mountain volcano? These rocks have a unique chemistry and are called **nepheline monzosyenite**. Never mind the long name, these rocks are eroded from House Mountain's magma chamber and gave rise to the final lavas that were erupted, a rock known as **basanitic nephelenite**.

The trail begins a long, uphill switchback towards the east. At the end of the switchback, follow the whitish soil off the trail for just a few steps and you'll be standing on the Ft. Apache Member of the Schnebly Hill formation. Look at the red hills around you and see if you can find this thin, gray limestone in outcrop at eye

level. This limestone marks
a moment in time when the
Pedregosa Sea extended its
reach into the Sedona area.
As you return to the trail
and continue up the last two
switchbacks, you'll begin to
notice grayish cinder material
giving way to red cinders
near the top of the mountain.
These deposits are the oldest
on House Mountain and
represent droplets of hot

Close-up of contact shown on page 136 (lens cap for scale)

lava that flew through the air and cooled to become cinders lying on the slope of the
ancestral rim. Before there was a shield volcano here, there was a cinder cone.

As the trail flattens and you start to see views to the south, you are looking across
the central crater of the volcano. Notice the two rounded hills on the floor of the
crater. These are called resurgent domes and they formed inside the crater after most
of the eruptions had ceased. This is a good place to have a snack or lunch and enjoy
the view to the north to the glorious red rocks in the distance. The "house" of House
Mountain is visible to the west of here and is a remnant lava flow that spilled from the
crest of the volcano towards the west. Old timers speak of a tree that once grew on
this remnant flow, giving the impression of a chimney rising out of a house from the
valley below, thus the origin of the name.

After a short rest here you can walk slightly uphill to the west on a very indistinct
social trail that hugs the inside wall of the crater. Do not walk directly towards the
"house" but rather more to its right as you angle upslope. Keep walking through the
shrubbery and you'll pick up a social trail. Once you've gone about 0.25 mile, you'll
begin to see red sandstone on your right in a small depression. At the place where the
black rock from House Mountain volcano meets the red rocks of the Schnebly Hill
Formation, you will be standing at the exact place where lava covered the slope of the
Mogollon Rim. Now, cross the low saddle and climb to the other side and up a small
white cliff composed of the Verde Formation. (Remember the white boulders you saw
on the trail below?) This is evidence that the Verde lake once completely covered the
House Mountain volcano, first in fresh lake water and then in deposits of limestone.
(I have found small remnants of white limestone on the "house.") Most of the lake
deposits are now eroded from the mountain but the evidence found here allows us to
imagine when lake water lapped on the flanks of the extinct volcano. See the chapter
"Creating the Modern Landscape" on page 91 for a more complete discussion of the
Verde lake. If you walk to the eastern edge of this small white mesa you will get a
great view of the Sedona area to the northeast.

Retrace your steps back to the red saddle, the red cinders, and the trail back down
to Turkey Tank and your vehicle. You'll be pleasantly tired at the end of this hike but
intellectually stimulated by the time traveling you have done.

GEOLOGY HIKE 2
The Saddle of Cathedral Rock

Cathedral Rock is one of Sedona's most photographed and recognizable landforms. Like many of Sedona's other well-known rocks – Coffee Pot Rock and Bell Rock to name just two – Cathedral Rock's mass is carved into the colorful Schnebly Hill Formation, a deposit that records a time when migrating sand dunes fringed the shore of an ancient sea. This short but steep hike will take you to the flat saddle between its two central spires, where you will get to stand on the remnants of an ancient sea that once washed across this part of Arizona. A volcanic dike can also be seen while standing in the saddle. There is no practical way to reach the very top of this monolith but the hike is exhilarating and the views from the saddle are spectacular, especially if you beat the crowds in the early AM or visit while others are rushing back for dinner.

Majestic Cathedral Rock

Access

From Uptown Sedona: From the roundabout connecting highways 89A and 179, drive south on 179 about 3.3 miles and turn right at the roundabout at Back O' Beyond Road.

From Village of Oak Creek: From the roundabout connecting Highway 179 and Bell Rock Road, drive north about 3.0 miles and turn left at the roundabout at Back O' Beyond Road.

From Highway 179: Turn west on Back O' Beyond Road and drive 0.7 miles to a parking lot located on the left side of the road. Display your Red Rock Pass before beginning the hike.

A Short, Easy Walk to a Great View

Before you arrive at the trailhead, you will drive through a small canyon within rocks belonging to the Supai Group (Esplanade Sandstone). These rocks are located on the up-thrown side of the Cathedral Rock fault and the parking area for the hike is built on top of the fault line. No need to worry though, as this fault is not considered active. A short, easy hike on USFS Trail #170 begins at the parking lot. Follow the rock "cairns" (cylindrical mounds of rock) across a small wash to the south and proceed about 0.3 miles until you come to a wide shelf of flat rock called Courthouse Butte Vista. The view from here is impressive and worth a long pause. The expansive view takes in Mitten Ridge to the northwest and Courthouse Butte and Bell Rock to the southeast. If climbing uphill steeply is not your thing, you can return to your car after enjoying the view from here.

A More Difficult Hike Up to the Saddle

After taking in the view from Courthouse Butte Vista, continue to follow the rock cairns an additional 0.4 miles uphill. This is a much more difficult stretch than the first part of the hike and some rudimentary climbing is involved. Just after starting this leg of the hike you will cross a hiking and biking trail called the Templeton Trail (signed) that traverses the base of Cathedral Rock. Beyond this the rock cairns jog to the right a bit and then continue uphill to the saddle. There are several steep sections along this part of the trail, which the Forest Service rates as difficult. Proceed carefully-there is a good reason that these sandstones are locally called "slickrock." You will huff and puff on this climb but after arriving at the saddle you'll be rewarded for your effort!

Cathedral Rock's Geology in Brief

The beginning of this hike starts upon the trace of the Cathedral Rock fault. Look back to the north and you will see the uplifted Supai Group. These same strata are faulted down beneath the trailhead about 300 feet. The hike starts within rocks of the Hermit Formation and along the way to Courthouse Butte Vista, the Hermit proceeds upward into the Schnebly Hill Formation. It is difficult to locate a sharp contact between the two but astute observers may note a slight color change in the soil from the darker, brick-red Hermit Formation to the lighter, orange-colored Schnebly Hill Formation. The Hermit is composed of siltstone and mudstone and is relatively soft and erodes easily. Soils develop readily on it, so when hiking through the trees at the start of the hike you are most likely walking on the Hermit Formation. When you exit the forest of trees and can see a broader expanse, you will know that you have passed the contact and into the harder, less forested Schnebly Hill Formation.

When the Hermit Formation was deposited some 285 Ma, ephemeral rivers flowed through the Sedona area and left silt, mud, and gravel on the arid floodplain. See the map on page 36 to get a better idea of how the landscape looked then. Through time, the environment became more arid and wind-blown sand entered the area from the northwest. The land to the southeast was subsiding at the same time and a shallow sea intruded from that direction. An interplay between wind-blown sand and marine environments allowed sand to become deposited along the shoreline. Tides and currents reworked the sand, moving it into horizontal beds known today as the Bell Rock Member of the Schnebly Hill Formation. Ultimately, and for a moment in time (which could be 500,000 years to a geologist), the sea transgressed fully into the Sedona area. It was at this time that the Ft. Apache Member was deposited as 10 feet of limestone and dolomite. This sea later regressed to the southeast and additional dune deposits covered the Ft. Apache Member. This overlying deposit is the Sycamore Pass Member.

As you climb uphill from Courthouse Butte Vista you pass through the entire Bell Rock Member. Note the flat-bedded nature of the sandstone, called slickrock, and interpreted as wind-blown sand reworked by the tides of the Pedregosa Sea. You will notice that the bulk of the deposit is red in color but here and there you may notice white-colored horizons. These, too, are part of the Bell Rock Member but are localized horizons where the red color has been leached out by groundwater long before the rocks were exposed by erosion. The reason these horizons are leached white is complex but likely involves the presence of finer-grained horizons below them that caused the groundwater to flow more abundantly above. The fine-grained sandstone below may be the result of the former presence of sabkha environments (salt flats) that existed on the desert floor some 280 Ma.

The "trail" is steep but eventually climbs to the gray, 10-foot ledge of limestone known as the Ft. Apache Member of the Schnebly Hill Formation. When the trail arcs back to the north you will see a large cave-like structure where the Ft. Apache Member is visible in the middle. This is a great place to rest or get out of the wind or sun, although the final few feet of the climb are nearby and tempting on calm or cooler days. The Ft. Apache Member was first named from outcrops near the famous fort located about 120 miles to the southeast. The unit pinches out entirely about five miles northwest of here. This is the information that geologists use to reconstruct the **paleogeography** for any moment in time. Thicker deposits mean that the sea was present for a longer time in that direction (but not that the sea was deeper as is commonly misconstrued).

The Ft. Apache Member exposed along the trail to the saddle

The Ft. Apache Member is what "holds up" the surface of the saddle between the two spires in Cathedral Rock. Rock enthusiasts usually must turn their gaze skyward to glimpse the Ft. Apache Member situated in the sheer wall of a cliff face. But at this spot you can touch it and walking on its broad promenade is a special treat. There are excellent views from the saddle to the southeast towards Courthouse Butte and Lee Mountain. Turning your gaze to the southwest, you'll see a mass of black rock emerging from the pinyon and juniper trees. This is a volcanic dike, an ancient pathway for magma that rose through the rocks here long before Cathedral Rock eroded into its iconic profile. The black

Lee Mountain (left) and Courthouse Butte (right) from Cathedral Rock saddle

rocks you see here were the last bit of magma that cooled and "froze" in place. The dike may be related to eruptions at the House Mountain volcano (15 to 13 Ma) or the plateau basalts (8 to 6 Ma).

If you choose, you can walk towards the west on a flat ledge of the Ft. Apache Member for another great view. Look up and you'll see the Sycamore Pass Member, which comprises the upper half of Cathedral Rock. As you look skyward to these rocks, you may notice the angled cross-beds that indicate how wind-blown sand dunes eventually covered the marine deposits of the Ft. Apache Member. These arid conditions prevailed for a few million years as northwest winds blew loose sand into the Sedona area. As the Pedregosa Sea progressively retreated to the southeast, dunes in the Sedona area changed their character slightly to that of inland dunes and became deposits known as the Coconino Sandstone. This interval is where red, cross-bedded sandstone (Schnebly Hill Formation) gives way to white, cross-bedded sandstone (Coconino Sandstone).

Stay here for a while and listen to the wind as it moves between these petrified dunes. The geology here is large, even to a geologist. In our everyday lives, we get all the encouragement we need to "do more" and "be more." But the lessons of geologic time remind us that there is a place in our lives for quiet contemplation and intelligent reflection. You are at such a place here on Cathedral Rock. Do not let this moment of beauty and learning escape you, by rushing off to some other place. Enjoy this moment!

DESCRIPTIONS of SEDONA'S ROCK LAYERS

The Supai Group

Age: 316 to 287 Ma
Periods: Pennsylvanian and Permian
Named: (Formation) 1910 by N.H. Darton, (Group) 1975 by Eddie McKee
Type Section: Near Supai Village, Cataract Creek, Grand Canyon
Maximum Thickness: 400 feet
Rock Types: Shale, sandstone, conglomerate, and minor limestone
Color: Red to soft red, buff, and gray
Structures: Eolian cross-bedding, cut-and-fill channels, structureless sandstone
Outcrop: Forms small cliff-and-slope (step-like) topography beneath Midgley Bridge and south of Airport Mesa
Environment: Fluvial floodplain and channels, eolian, shallow marine
Fossils: None
Formations: Watahomogi, Manakacha, Wescogome, Esplanade Sandstone
Correlation: Wescogame Formation: Honaker Trail Formation (Utah); Esplanade Sandstone: Cedar Mesa Sandstone and Cutler Group (Utah)

History: Rocks known today as the Supai Group were originally part of the Aubrey Group, named in 1875 by famed geologist G. K. Gilbert. In 1910, N. H. Darton subdivided the lower reddish parts of this group into the Supai Formation, which included rocks of the Hermit Shale. In 1923, Levi Noble removed the Hermit Shale from the Supai Formation and added some beds formerly belonging to the Redwall Limestone (today's Watahomogi Formation). In 1975, Eddie McKee raised the section to group status that included four formations. These formations are difficult to discern and some advocate a return to formation status (with members) for the Supai.

Geology: The Supai represents a great transition in the ancient environments of the American Southwest. For 200 million years prior to this time, the area was the site of shallow marine settings. This environment waned as Supai deposition commenced with thin beds of limestone and dolomite interbedded with increasing thicknesses of siltstone and sandstone. By Esplanade Sandstone time, widespread eolian conditions existed. Continental environments would dominate for the next 200 million years in the Southwest. The Supai is composed dominantly of inter-bedded channel sandstone and conglomerate with floodplain mudstone and siltstone. Geologists interpret this to mean that river channels shifted laterally across the rather featureless floodplain. Eolian sandstone increases up-section and documents a trend towards more arid conditions. The Persian Gulf region may be a good modern analog.

The Hermit Formation (Shale)

Age: 285 to 280 Ma
Period: Permian
Named: (Shale) 1923 by Levi Noble; (Formation) 1990 by Ron Blakey
Type Section: Hermit Basin, Grand Canyon
Maximum Thickness: 300 feet
Rock Types: Mudstone, siltstone, sandstone, and conglomerate
Color: Dull to brick-red
Structures: Cut-and-fill channels, mud cracks, ripple marks, root casts
Fossils: *Walchia* (primitive conifer)
Outcrop: Forms broad, tree-covered slopes and terraces that are often poorly exposed in Sedona; in roadcuts west of the post office
Environment: Arid, fluvial floodplain and ephemeral river channels
Correlation: Organ Rock Shale (Utah); Cutler Group (Colorado); Abo Formation (New Mexico)

History: The Hermit Formation was originally described for exposures in the Grand Canyon along the Hermit Trail. It was initially considered part of the Supai Formation but was elevated to formation status by Levi Noble in 1923. He named it the Hermit Shale and references to this specific name will often be found in the historic literature. It was informally renamed the Hermit formation by A. H. McNair in 1951, and this scheme was formalized in 1990 by Ron Blakey of Northern Arizona University, who noted the obvious lack of shale lithologies in the Mogollon Rim area.

Geology: The Hermit Formation forms the broad terrace upon which most of the town of Sedona is built. These easily weathered siltstones and mudstones weather easily and this helps create the bench-like topography. Hermit deposition continued an overall arid fluvial setting inherited from Supai time. Rivers emanating from the Ancestral Rocky Mountains in southwest Colorado flowed southwest towards the Sedona area and eventually went to the sea in the Las Vegas/St. George area. The Hermit is correlated with the Organ Rock Formation in southeast Utah and the Cutler Group in southwest Colorado, where the Ancestral Rocky Mountain front was located. Fossils of Walchia (a primitive conifer) suggest an arid environment but riparian conditions may have been present along stream courses. The Persian Gulf region may be an appropriate modern analog.

The Schnebly Hill Formation

Age: 280 to 275 Ma
Period: Permian
Named: 1990 by Ron Blakey
Type Section: Schnebly Hill east of Sedona
Maximum Thickness: 700 feet
Rock Types: Sandstone, siltstone, mudstone, minor limestone, and dolomite
Color: Orange to brick-red, thin limestone lenses are gray and white
Structures: Eolian cross-bedding, wind ripples, tidal ripples, adhesion ripples
Fossils: Trackways of reptiles on some cross-beds
Members: Rancho Rojo, Bell Rock, Ft. Apache, and Sycamore Pass
Outcrop: Forms steep stair-step (lower part) to vertical (upper part) cliff faces everywhere above Sedona in the form of buttes and the Mogollon Rim
Environment: Eolian coastal desert (some reworked by the tides)
Correlation: De Chelly Sandstone (Arizona)

History: In 1945, Eddie McKee originally assigned rocks of the Schnebly Hill Formation to part of the Supai Formation. Ron Blakey began studying these rocks in detail in 1979, and discovered that they belonged to a unit not present in the Grand Canyon area. In 1990, he formalized the name Schnebly Hill Formation, having four members. Only the rare geologist still disputes whether these rocks are separate and distinct from the Supai Group. The debate has quieted down somewhat from the "war years" of the 1980s, and most geologists accept the classification scheme suggested by Dr. Blakey.

Geology: The Schnebly Hill Formation represents a time when dunes marched into the Sedona area from the northwest. At the same time, a shallow arm of the Pedregosa Sea began inundating Arizona from the southeast. The Rancho Rojo Member is an initial coastal dune deposit. Later, the coastal dunes were reworked by the tides of the Pedregosa Sea, leaving flat-bedded and ripple-laminated sandstone exposed in the Bell Rock Member. The Pedregosa Sea eventually reached the Sedona area for a brief moment in time and the marine Ft. Apache Member (limestone and dolomite) was laid down. This gray ledge, about eight to ten feet thick in the Sedona area, is an excellent marker bed in a "sea" of orange sandstone. As the sea retreated to the southeast, cross-bedded sand laid down in huge sweeping dunes, covered the landscape with the Sycamore Pass Member. Many well-known landmarks in Sedona (Bell Rock, Coffee Pot Rock, and Cathedral Rock) are the eroded remnants of these ancient dune and tidal landscapes. The Namibian coastline in southwest Africa may be a good modern analog.

The Coconino Sandstone

Age: 275 to 273 Ma
Period: Permian
Named: 1910 by N. H. Darton
Type Section: Coconino Plateau, south of Grand Canyon
Maximum Thickness: 560 feet
Rock Type: Sandstone
Color: Yellow, gold, buff, white, gray
Structures: Eolian cross-bedding, minor rain drop impressions
Outcrop: Forms steep, light-colored cliffs with eroded (slanted) cross-bed surfaces everywhere above Sedona on buttes and the Mogollon Rim
Environment: Eolian, sandy, inland desert
Fossils: Trackways of reptiles in cross-beds
Correlation: Glorieta Sandstone (New Mexico)

History: The Coconino Sandstone was originally a part of the Aubrey Group named by G. K. Gilbert in 1875. N. H. Darton gave it a formal name in 1910, separating it from the Supai below and the Kaibab above. Unlike other rock units in the area, the Coconino has kept its original name with very little tampering by later geologists. That is because it is a readily defined package of cross-bedded eolian sandstone. In 1990, Ron Blakey proposed two members for the Coconino Sandstone in Oak Creek Canyon, however, this classification is not recognized outside of the canyon area and the nomenclature has not taken hold.

Geology: The Coconino Sandstone represents one of the great fossil deserts of all time. As the Permian winds blew from the north, material that originally had been transported by rivers from the mighty Appalachians and the Ancestral Rocky Mountains, was then blown to the south forming huge desert dunes. As the dunes migrated south they left evidence of the ancient wind direction and this is recorded in the southern dip of their cross-beds. As the land slowly subsided space was created that put the dunes into silent preservation. Occasionally, reptiles walked the dunes and left their foot or tail impressions. Today the Coconino is a regional aquifer and many springs originate within the unit in the upper reaches of Oak Creek Canyon. A good modern analog is the sand seas of the Arabian Peninsula or the Sahara Desert in Africa.

The Toroweap Formation

Age: 273 to 272 Ma
Period: Permian
Named: 1938 by Eddie McKee
Type Section: Toroweap Valley, north side of Grand Canyon
Maximum Thickness: 240 feet
Rock Types: Siltstone, sandstone, carbonate, evaporite
Color: White to gray to reddish-pink
Structures: Crinkly (deformed) bedding, sabkha textures, cross-bedding
Fossils: Brachiopods, corals, bryozoans (in western outcrops)
Outcrop: Forms slopes of easily eroded material west of Sedona; resistant sandstone cliffs in Oak Creek Canyon
Environment: Sabkha (salty tidal flat), beach, coastal dunes
Correlation: White Rim Sandstone (Utah)

History: The Toroweap Formation was named and described by Eddie McKee for exposures in the spectacular Toroweap Valley in western Grand Canyon. It was originally part of the Aubrey Group (Gilbert, 1875), then part of the Kaibab Limestone (Darton, 1910). In 1938 McKee broke it out of the Kaibab and gave it its own name. In the Grand Canyon area it is composed of three members but no such differentiation is noted in the Mogollon Rim. However, Blakey calls the Toroweap Formation in Oak Creek Canyon the Cave Springs Member of the Coconino Sandstone.

Geology: The Toroweap is a fascinating deposit that undergoes a rapid facies change (change in character or rock type) in a five-to-ten mile distance along the Mogollon Rim in Sedona. To the west in Sycamore Canyon, it is composed of soft siltstone and evaporite deposited on a tidal flat that was occasionally exposed to the coastal winds. Gypsum allows this part of the Toroweap to erode easily, forming sloped topography. Gradually, almost imperceptibly, the Toroweap changes to a well-consolidated, cross-bedded sandstone towards the east. Geologists interpret this to mean that this section of the unit was laid down in dunes in the onshore environment. This fantastic facies change documents the shoreline transition zone between a shallow marine sabkha to the west and coastal dunes to the east. The transition zone is also found in the subsurface towards the Marble Canyon area near Lees Ferry. The coastal dunes in Namibia may be modern equivalent,

The Kaibab Formation (Limestone)

Age: 272 to 270 Ma
Period: Permian
Named: (Limestone) 1910 by N.H. Darton; (Formation) 1938 by Eddie McKee
Type Section: Kaibab Plateau on the north side of Grand Canyon
Maximum Thickness: 180 feet
Rock Types: Limestone, sandy dolomite, chert
Color: White to gray
Structures: Tabular-bedded and ripple-laminated limestone
Fossils: Brachiopods, corals, sponges, bryozoans, shark teeth
Outcrop: Forms ragged to craggy cliffs with steep slopes
Environment: Shallow marine
Correlation: Phosphoria Formation (Utah); San Andres Limestone (New Mexico)

History: Named originally as the uppermost part of the Aubrey Group by Gilbert in 1875, then subdivided out in 1910 as the Kaibab Limestone by N. H. Darton. The underlying Toroweap Formation was separated from the Kaibab Limestone by Dr. E. D. McKee in 1938, who was the first to formally call it the Kaibab Formation. Later references continued to call it the Kaibab Limestone and this tradition persists in everyday usage at Grand Canyon. Ralph Lee Hopkins completed a detailed Master's thesis on the unit in 1986, using the term "Formation." The Kaibab is variously called limestone or formation.

Geology: The Kaibab Formation is the cap rock for most of the Grand Canyon and the southern Colorado Plateau. The relatively large percentage of silicious chert in the limestone makes it a durable unit that resists weathering in the modern arid environment. The chert may have had its origins in the profusion of sponges that inhabited the Kaibab Sea. Their spicules rained down on the sea bed as a source of silica. The Kaibab represents a greater extension of the marginal marine conditions of the Toroweap Formation. However, just east of Sedona in the subsurface, the Kaibab undergoes a facies change similar to that of the Toroweap, turning sandy and eventually disappearing altogether in eastern Arizona. It's safe to say that without the Kaibab serving as a cap rock in this part of the world, the landscape might easily be more denuded than what is seen today. Any modern shallow sea would serve as an analog for the Kaibab.

Rim gravel

Age: 37 to 30 Ma
Period: Paleogene
Named: An informal name first proposed by M. E. Cooley in 1963
Type Section: On top of the Mogollon Rim
Maximum Thickness: Maximum 180 feet
Rock Types: Conglomerate and sandstone
Color: Red to reddish- gray
Structures: Flat-bedded to trough-bedded channel gravel and sand
Fossils: None
Outcrop: Very indistinct and found in extremely isolated outcrops on top of the Mogollon Rim. Closest exposure to Sedona may be on top of the rim near Sycamore Canyon.
Environment: Fluvial
Correlation: Mogollon Rim Formation (Arizona), Claron Formation (Utah)

History: Deposits of these gravels are inconspicuous and widely separated so it's not surprising that they have never been formally described. M. E. Cooley traced the history of the Mogollon Highlands from Triassic to Cenozoic time, and in turn was forced to name it. Don Elston suggested that the Rim gravel was once a much thicker deposit that completely buried the Mogollon Rim (and the Grand Canyon for that matter) before being stripped to its present isolated outcrop pattern. Paul Lindberg argues that gravel deposits on the valley floor near Sedona are merely the down-faulted portions of the Rim gravel, placed there by movement on faults that helped create the Verde Valley. Andre Potochnik named the Mogollon Rim Formation for similar rocks on the White Mountain Apache Indian Reservation in eastern Arizona and believes that they are correlative with the Rim gravel.

Geology: Perhaps no other deposit on the Colorado Plateau has generated so much interpretation from so little rock. Only a few isolated exposures of the Rim gravel are present on top of the rim, yet they reveal that uplift and erosion was occurring to the south of Sedona between about 45 and 30 Ma. The deposit has many Precambrian-age clasts and since those rocks are only exposed to the south that is the direction they must have come from. Most likely, the Rim gravel represents deposition during a time when the landscape was not yet dissected vertically. Broad floodplains likely stretched from central Arizona to the center of the Colorado Plateau. Subsequent erosion has reduced the extent of the Rim gravel to isolated outcrops only. A modern analog might be the gravely sections of the Great Plains east of the Rocky Mountains.

Beavertail Butte formation

Age: 25 to 20 Ma
Period: Late Paleogene and early Neogene
Named: An informal name first proposed by Wayne Ranney in 1988
Type Section: Beavertail Butte south of Sedona
Maximum Thickness: 200 feet
Rock Types: Conglomerate, sandstone, siltstone, and marl
Color: Reddish-gray to pink
Structures: Flat-bedded, trough-bedded, and crudely stratified
Fossils: None
Outcrop: Forms low, rounded hills when exposed to erosion, cliff faces when capped by harder units
Environment: Fluvial and alluvial fan with local lacustrine settings
Correlation: None

History: Mahard (1949) was the first to map and recognize sedimentary rocks that he lumped within the Hickey Formation. Twenter and Metzger (1963) followed up and described the lithology and interpreted their source area. Peirce and Nations (1979) showed that these deposits were likely separate and distinct from the Rim gravel based on their substrate. Ranney (1988) mapped the deposit in the Sedona and House Mountain areas and proposed the name. Loeske (1999) used this nomenclature in a detailed study of the Beavertail Butte formations relationship to regional tectonics. Richard Holm (2001) made a detailed study of the Cenozoic sedimentary rocks in north-central Arizona and documented that the Beavertail Butte formation is a separate and distinct deposit from the similar looking Rim gravel.

Geology: The Beavertail Butte formation is in many ways a continuation of the environments from Rim gravel time, with an important exception. Rivers continued to shed material from the south in the Mogollon Highlands but by Beavertail Butte time, these gravels were blocked from further northeast travel by the presence of the Mogollon Rim. In the area south of the Village of Oak Creek these gravels are up to 200 feet thick but pinch out abruptly to the northeast against a rising slope of Paleozoic strata-evidence for the early Miocene existence of the escarpment. The unit was divided by Ranney (1988) and Loeske (1999) into three members, a lower member with localized, angular clasts from the ancestral rim, a middle member with marl and mudstone, and an upper member with clasts derived in fluvial environments from the south.

Hickey Formation

Age: 15 to 10 Ma
Period: Neogene
Named: 1958 by Anderson and Creasey
Type Section: Hickey Mountain in the Black Hills above Jerome
Maximum Thickness: 1,400 feet
Rock Types: Mostly basalt flows, cinders, and dikes; minor andesite
Color: Brown-black to gray
Structures: Massive flows, weathered cinders, basalt columns, dikes
Fossils: None
Outcrop: Forms cliffs with very steep slopes unless weathered
Environment: Volcanic vents yielding cinder cones and shields
Correlation: None

History: Anderson and Creasey (1958) first proposed the name Hickey Formation for "mediocre exposures" of volcanic and sedimentary rocks on Hickey Mountain, northwest of Mingus Mountain in the Black Hills. They stated that a more typical section was located on Mingus Mountain and logically wanted to use that name, but it was assigned to another unit. Anderson and Creasey did not recognize that the House Mountain basalt was related temporally to the lava on Mingus Mountain and originally included the Beavertail Butte formation with the Hickey. Twenter and Metzger (1963) later separated out the sedimentary rocks.

Geology: The Hickey Formation represents a time (generally between 15 and 10 Ma) when basaltic lava poured out effusively over the central Arizona landscape. The deposit extends from the northern fringes of the Salt River basin (the Cave Creek area near Phoenix) to the House Mountain volcano, although minor deposits could be buried by younger lavas to the north of this. When the lavas were emplaced on Mingus Mountain, the Verde fault had not yet ruptured and an undulating but southward-rising surface lay beneath the volcanic field from the base of the ancestral Mogollon Rim (Sedona) to the mostly eroded foothills at the base of the Mogollon Highlands (Prescott). Near the type section on Mingus Mountain, the Hickey Formation is composed of basalt lava flows 1,400 feet thick. Recently, the rocks in the Verde Valley west of House Mountain have been studied and they show the same unique chemistries that characterize the latest flows on House Mountain.

Verde Formation

Age: 9 to 3 Ma
Period: Neogene
Named: 1923 by O. P. Jenkins
Type Section: Verde Valley and areas near the Verde River
Maximum Thickness: 3,300 feet
Rock Types: Impure limestone, mudstone, and evaporites (salt and gypsum)
Color: White to light gray limestone, brown mudstone
Structures: Massive to crinkly-bedded limestone and structureless mudstone
Fossils: Stegomastodon, snails, plant roots, and the trackways of camels, canides, and other mega-fauna
Outcrop: Forms small stair-step cliff and slope topography
Environment: Lacustrine, fresh water lake and playa
Correlation: Deposits in other closed basins in Arizona

History: The Verde Formation was first named and described by O. P. Jenkins in 1923. (It's incredible to think that these rocks received a formal name before the red rocks in Sedona did.) Dr. Dale Nations did a detailed study of its stratigraphy in 1981. Paleomagnetic dates on the rocks were obtained in 1978 by Bressler and Butler.

Geology: The Verde Formation represents the time when the Verde Valley attained its modern shape, occasionally holding lake water during certain climate and tectonic regimes. As the Verde fault became active about 10 Ma, the Verde River was trapped between the uplifted Black Hills near Jerome and the eroded and partially faulted edge of the Mogollon Rim near Sedona. Continued fault movement caused the Verde Valley to drop and the river was trapped occasionally in lacustrine or lake environments. Evaporites and inter-bedded mudstone show that the lake would occasionally dry up and become an expansive playa environment. It was along the shores of this fluctuating lake that many large mammals walked, leaving their trackways and occasionally a fossil in the limestone. At one time lake deposits completely buried the House Mountain volcano. Dating of the rocks suggests that the Verde River became free-flowing again between about 2.5 to 3.0 Ma, leaving the white, rounded hills in the floor of the modern valley.

Glossary

adhesion ripples: a sedimentary structure consisting of an irregular or blistered sand surface, formed by the wind blowing dry sand over a moist surface

alluvium: a general term for all sedimentary deposits laid down by rivers, especially at the foot of mountain ranges where deposits form alluvial fans

Ancestral Rocky Mountains: a mountain range present in southwestern Colorado and adjacent areas from about 320 Ma to 165 Ma

aquifer: stratum or zone below the surface of the earth capable of holding and producing water as in a spring or well

arroyo: the channel of an intermittent stream, usually with steep sides

basalt: generally a dark, fine-grained volcanic rock with about 52% silica content; the most common rock type on the surface of planet

basanitic nephelenite: a rock similar in outward appearance to basalt but which chemically is rich in the plagioclase minerals and strongly undersaturated in feldspathoid minerals.

bedrock: referring to any consolidated rock that underlies the rock or sediment being described; generally used to refer to older rocks

beveled: the planning by erosion of the edges of strata

caliche: a desert soil formed by the near-surface crystallization of calcite or other soluble minerals by upward-moving solutions

Cenozoic: the latest of the four geologic eras into which geologic time is divided; it spans from about 65 Ma to the present.

chert: any microcrystalline variety of quartz; red chert is called jasper and black chert is called flint

clast: an individual fragment or grain in a sedimentary rock (i.e. clay, sand, pebbles, etc.) formed by the physical disintegration of some larger rock

Colorado Plateau: the region of the Four Corner states (Arizona, Colorado, New Mexico, and Utah) that was uplifted gradually, escaping the more intense deformation of surrounding areas. It is one of three geologic provinces in Arizona and is characterized by high plateaus and deep canyons.

colluvium: loose, unconsolidated deposits brought down by gravity and found at the foot of a cliff or slope; colloquially called talus

compression: a system of forces or stresses that tends to decrease the volume of or shorten a substance; the squeezing of rock layers in mountain belts

conglomerate: rounded, water-worn fragments of rocks or pebbles cemented together by another mineral substance

contrary bedding: strata that appear to have been tilted opposite to the sense of displacement usually formed in a previous faulting event

contact: the place or surface where two different types of rocks come together. It is found at the top of one formation and the base of the overlying one.

correlation: the determination of age equivalence in two geologic strata or other event

cross-bedding: the arrangement of laminations of strata at an angle to the main planes of horizontal stratification. Cross-bedding is useful in determining the source direction of wind or water currents.

depositional environment: the surface conditions, such as climate, geographic setting, and transport medium that affect the nature of sedimentary deposits

dike: a tabular, or flat body of igneous rock (magma) that cuts across the structure of adjacent rocks

dolomite: a rock similar to limestone but containing more magnesium

drainage: any area where water is removed by down-slope flow. Drainages range in scale from large rivers to small rivulets.

dune: a low hill or bank of drifted or wind-blown sand

en echelon: parallel structural features that are offset

eolian: from Eolus, the Roman god of wind; refers to sediment that is transported and deposited by the wind

ephemeral: in geology, pertaining to conditions that change, such as normally dry drainages that carry water only after storms

entrain: to attach to or get caught up with

erg: a large sand "sea"

escarpment: the steep face frequently formed by the abrupt erosion or termination of sedimentary rocks; a colloquial term is cliff.

evaporite: sediments which are deposited from aqueous solution as a result of extensive evaporation of the solvent

extension: stress caused in the earth's crust by pulling apart, as opposed to compression

facies: general appearance and make-up of a rock body as contrasted with other parts by its texture and composition. Different environmental conditions from one place to another will cause sediments to undergo a facies change.

fault: a fracture or zone of fractures within the earth's crust caused by compression or tension stresses. Faults show where rocks have been displaced.

fissure: a break or fracture where magma rises within the earth and extrudes lava onto the surface

floodplain: that portion of a river valley that is only submerged when the river overflows its banks (or levee) during flooding

fluvial: pertaining to rivers or river environments

formation: the primary unit or division of sedimentary rocks. Formations possess certain distinctive features such as rock type, bedding characteristics, or age.

gaining reach: that portion of a stream in which the volume of water increases

graben: a fault-bounded valley down-dropped between two uplifted blocks

groundwater sapping: the geomorphic process in which groundwater exits a hillslope laterally in seeps and springs, and erodes soil or rock from the slope. This often causes the slope to be undermined and undergo mass wasting.

gradational contact: when two different rock types intertongue with one another, representing a transition phase between the deposits

group: a stratigraphic unit consisting of two or more related formations

gypsum: a calcium sulfate mineral formed when saline water evaporates

headwall: the steep, wall-like cliff at the upper end of a stream or drainage

headward erosion: a process whereby a valley or canyon is lengthened and widened at its upper end by the water that flows at its head

horizon: a particular level or bed within a sediment, with or without thickness

index fossil: a fossil characteristic of an assemblage zone and restricted to it

interstices: pore or void

inverted topography: a feature formed when any hard rock originating in a low, confined valley, is subsequently left standing high when the softer valley walls erode away

lacustrine: pertaining to lakes or lake environments.

limestone: a bedded sedimentary deposit composed mainly of calcium

losing reach: that portion of a stream in which the volume of water decreases due to infiltration into the ground

lithology: the physical character of a rock

marine: pertaining to sea or ocean environments; not synonymous with fresh water environments

matrix: In a rock in which some grains or clasts are larger than others, the smaller-size material is the matrix.

mature: sediment in which all, or nearly all, of the easily weathered material is absent

member: a subdivision of a formation differentiated by rock type or bedding characteristics; usually not thick enough to be considered its own formation

Mesozoic: the third era of geologic time ranging from about 251 Ma to 65 Ma; also known as the Age of Dinosaurs

Mogollon Highlands: a mountain range present in central Arizona from about 80 Ma to 17 Ma

Mogollon Rim: an elongate escarpment or cliff face in central Arizona whose western end comprises the towering cliffs or red sandstone near Sedona

mudstone: a common type of sedimentary rock, composed of a mixture of silt and clay-size clasts. Mudstone with regular, thin banding is called shale.

Ma: abbreviation for mega-annum, meaning millions of years ago.

nepheline monzosyenite: an intrusive rock composed of grains of alkalai feldspar, nepheline, and an alkalic ferromagnesian constituent

nearshore: pertaining to that part of the environment close to the shoreline and influenced by wave action

nomenclature: the names used in any classification system

onlap: the extension or protrusion of successive layers of sediment beyond the marginal limit of their predecessors

orogeny: the process of mountain building or formation, especially by the folding of rock layers

outcrop: the exposure of bedrock or strata from beneath its overlying cover or soil

paleoflow: the ancient direction of flow for a river or lava

paleogeography: the study of ancient geography or landforms

paleosol: a soil from the past; a buried soil

Paleozoic: the second era of geologic time lasting from about 542 Ma to 251 Ma

Pennsylvanian: a division of time (Period) within the Paleozoic Era lasting from about 318 Ma to 299 Ma

Permian: the final Period of time within the Paleozoic Era lasting from about 299 Ma to 251 Ma

point bar: sediment deposited on the inside curve of a growing meander loop

porosity: the ratio between clasts and the total volume of a rock body

Precambrian: the first, and by far the longest, era of geologic time covering eight-ninths of all Earth history from about 4550 Ma to 542 Ma

province: a large area or region unified in some way and considered as a whole

radiometric dating: the technique which measures the ratio of decayed radioactive gases between parent and daughter products

reentrant: any indentation in a landform usually angular in character

relief: the net change in elevation from one locality to the next

root casts: the visual or replaced remains of plant roots in a rock

residence time: the amount of time that groundwater has been held in the subsurface

retreat(ed): in geology, referring to the erosion of a cliff or escarpment away from a highland

sabkha: An Arabic word that describes low coastal deserts, commonly covered with evaporite minerals such as gypsum or salt

sandstone: a cemented and compacted sedimentary rock usually composed of quartz grains; typically bedded or stratified but sometimes without bedding

seaway: a large body of saltwater usually smaller than an ocean

shield volcano: a broad, gently sloping (usually 15 degrees or less) volcanic cone composed mostly of basalt lava

slickensides: a polished and striated surface that results from friction along faults

slickrock: a term referring to the tendency of sand grains within sandstone to become loosened on exposed surfaces, causing one's feet to slip; generally applied to any exposed sandstone surface in the American Southwest, whether it is "slick" or not

sorting: a descriptive term used to indicate the degree of similarity in a sediment, with respect to the size of its grains

strata: the layering found within sedimentary rocks, regardless of thickness

stratigraphy: the branch of geology that deals with the study of strata or sedimentary rocks

subaerial: formed, existing, or taking place on land, as opposed to subaqueous

subaqueous: existing, formed, or taking place in or under water

submarine: existing, formed, or taking place in or under seawater; distinct from subaqueous

texture: the geometric aspects of the particles of a rock including size, shape, and arrangement

transition zone: one of the three geologic provinces in Arizona; it displays characteristics of both the Colorado Plateau and the Basin and Range provinces

trend: the direction or bearing of the outcrop of a bed or dike

tuff: a rock formed of compacted volcanic fragments

type section: a geographic area from which a rock formation receives its name

unconformable: not succeeding the underlying strata in immediate order of age

unconformity: a surface of erosion that represents a gap in the rock record

unroofed: the sequential removal by erosion of rock from a mountain range resulting in an inverse age of clasts in the outwash deposits

References

Anderson, C. A., and Creasey, S. C. 1958. Geology and ore deposits of the Jerome area, Yavapai County, Arizona. U.S. Geological Survey Professional Paper 308.

Blakey, R. C., and R. Knepp. 1989. Pennsylvanian and Permian geology of Arizona, in Jenney, J. P., and S. J. Reynolds, eds. Geologic evolution of Arizona: Arizona Geological Society Digest 17:313-47.

Blakey, Ronald C. 1990. Stratigraphy and geologic history of Pennsylvanian and Permian rocks, Mogollon Rim region, central Arizona and vicinity. Geological Society of America Bulletin 102:1189-1217.

Blakey, Ronald, and Wayne Ranney. 2008. Ancient Landscapes of the Colorado Plateau. Grand Canyon Association.

Cooley, M. E., and E. S. Davidson. 1963. The Mogollon Highlands: Their influence on Mesozoic and Cenozoic erosion and sedimentation. Arizona Geological Society Digest 6:7-35.

Duffield, J. A. 1985. Depositional Environment of the Hermit Formation, central Arizona. (Master's Thesis, Northern Arizona University, Flagstaff, Arizona.)

Elston, D. P., E. D. McKee, G. R. Scott, and G. D. Gray. 1974. Miocene-Pliocene volcanism in the Hackberry Mountain area and evolution of the Verde Valley, north central Arizona, in Karlstrom, T. N. V., Swan, G. A., and Eastwood, R. L., eds. Geology of northern Arizona, Part II, Area studies and field guide: Geological Society of America Rocky Mountain Section meeting guidebook, Flagstaff: Northern Arizona University.

Holm, Richard F., and Robert A. Cloud. 1990. Regional significance of recurrent faulting and intracanyon volcanism at Oak Creek Canyon, southern Colorado Plateau, Arizona. Geology 18:1014-17.

Holm, Richard F., and James H. Wittke. 1996. Geologic map of the summit area of House Mountain, Yavapai County, Arizona. Arizona Geological Survey Contributed Map CM-96-C.

Holm, Richard F. 2001. Cenozoic paleogeography of the central Mogollon Rim, southern Colorado Plateau region, Arizona, revealed by Tertiary gravel deposits, Oligocene to Pleistocene lava flows, and incised streams. Geological Society of America Bulletin 113:1467-85.

Jenkins, O. P. 1923. Verde River lake beds near Clarkdale, Arizona. American Journal of Science, 5th series 5:65-81.

Levings, Gary W. 1980. Water resources in the Sedona area, Yavapai and Coconino counties, Arizona. Arizona Water Commission, Bulletin 11.

Lindberg, Paul A. 1983. Development of the Verde graben, north central Arizona. [abstract] Arizona-Nevada Academy of Science, 27th Annual Meeting Flagstaff, Arizona,

Lindberg, Paul A. 2008. Subsurface groundwater flow and sinkhole development; Sedona, Arizona. AIPG Field Trip Guidebook. Northern Arizona University, Flagstaff, Arizona.

McAllen, W. R. 1984. Petrology and diagenesis of the Esplanade Sandstone, central Arizona. (Master's Thesis, Northern Arizona University, Flagstaff, Arizona.)

McKee, E. D. 1938. The environment and history of the Toroweap and Kaibab formations of northern Arizona and southern Utah. Publication 492. Washington: Carnegie Institution.

———. 1945. Oak Creek Canyon. Plateau Magazine 18: 25-32.

———, and E. H. McKee. 1972. Pliocene uplift of the Grand Canyon region: Time of drainage adjustment. Geological Society of America Bulletin 83:1923-32.

Nations, J. D., R. H. Hevly, D. W. Blinn, and J. J. Landye. 1981. Paleontology, paleoecology, and depositional history of the Miocene-Pliocene Verde Formation, Yavapai County, Arizona. Arizona Geological Society Guidebook 13:133-149.

Peirce, H. W., P. E. Damon, and M. Shafiqullah. 1979. An Oligocene (?) Colorado Plateau edge in Arizona. Tectonophysics 61:1-14.

Potochnik, A. R. 1989. Depositional style and tectonic implications of the Mogollon Rim Formation (Eocene), east-central Arizona. New Mexico Geological Society Guidebook. 40th Field Conference, Southeastern Colorado Plateau.

Ranney, Wayne D. R. 1988. Geologic history of the House Mountain area, Yavapai County, Arizona. (Master's Thesis, Northern Arizona University, Flagstaff, Arizona.)

Rawson, Richard R., and Christine E. Turner-Peterson. 1979. Marine-carbonate, sabkha, and eolian facies transitions within the Permian Toroweap Formation, Northern Arizona, in Baars, D. L., ed., Permianland: Four Corners Geological Society, Field Conference, 9th Guidebook:87-99.

Twenter, F. R., and D. G. Metzger. 1963. Geology and ground water in Verde Valley, the Mogollon Rim region, Arizona. U.S. Geological Survey Bulletin 1177.

Wittke, James H., and Richard F. Holm. 1996. The association — basanitic nephelenite — feldspar ijolite — nepheline monzosyenite at House Mountain volcano, north-central Arizona. The Canadian Mineralogist. 34:221-240.

Index

About the Author

Wayne Ranney is a geologist, trail guide, and author who became interested in earth history in 1975 while working at Phantom Ranch at the bottom of the Grand Canyon. Since that time, his life has revolved around the canyons, rivers, and red-rock stratigraphy found in the American Southwest. He received both his bachelor's and master's degrees from Northern Arizona University in his hometown of Flagstaff, Arizona. Wayne completed a detailed map of the House Mountain area, which showed that the volcano was three times older than previously thought and had erupted upon a pre-existing topography, possibly the ancestral Mogollon Rim. This study won the Eddie McKee Award at the Museum of Northern Arizona in 1988.

Wayne is a semi-retired professor who ocasionally teaches courses at Northern Arizona University in Flagstaff. He leads field trips throughout the American Southwest for the Museum of Northern Arizona and the Grand Canyon Field Institute. He is a scholar and lecturer on a variety of international expeditions and has visited and lectured on all seven of the earth's continents, including assignments at both the North and South poles. His other books include *Carving Grand Canyon*, *Ancient Landscapes of the Colorado Plateau*, and *Images: Jack Dykinga's Grand Canyon*. Wayne's current research and educational interests include studies of how and when the Colorado River evolved, writing geology articles and books for interested non-specialists, and leading educational excursions to many of our planet's most interesting landscapes. He lives with his wife Helen in Flagstaff. You can visit his web site at: **www.wayneranney.com**, and his geology blog at: **earthly-musings.blogspot.com**.